Multiplicity diagrams can be viewed as schemes for describing the phenomenon of "symmetry breaking" in quantum physics: Suppose the state space of a quantum mechanical system is a Hilbert space V, on which the symmetry group G of the system acts irreducibly. How does this Hilbert space break up when G gets replaced by a smaller symmetry group H? In the case where H is a maximal torus of a compact group, a convenient way to record the multiplicity is as integers drawn on the weight lattice of H.

The subject of this monograph is the multiplicity diagrams associated with $U(n)$, $O(n)$, and the other classical groups. It presents such topics as asymptotic distributions of multiplicities, hierarchical patterns in multiplicity diagrams, lacunae, and the multiplicity diagrams of the rank-2 and rank-3 groups. The authors take a novel approach, using the techniques of symplectic geometry. They develop in detail some themes that were touched on in *Symplectic Techniques in Physics* (V. Guillemin and S. Sternberg, Cambridge University Press, 1984), including the geometry of the moment map, the Duistermaat–Heckman theorem, the interplay between coadjoint orbits and representation theory, and quantization.

T0292350

SYMPLECTIC FIBRATIONS AND MULTIPLICITY DIAGRAMS

SYMPLECTIC FIBRATIONS AND MULTIPLICITY DIAGRAMS

VICTOR GUILLEMIN   EUGENE LERMAN
Massachusetts Institute of Technology

SHLOMO STERNBERG
Harvard University

CAMBRIDGE
UNIVERSITY PRESS

CAMBRIDGE UNIVERSITY PRESS
Cambridge, New York, Melbourne, Madrid, Cape Town, Singapore, São Paulo, Delhi

Cambridge University Press
The Edinburgh Building, Cambridge CB2 8RU, UK

Published in the United States of America by Cambridge University Press, New York

www.cambridge.org
Information on this title: www.cambridge.org/9780521111867

First published 1996
This digitally printed version 2009

A catalogue record for this publication is available from the British Library

Library of Congress Cataloguing in Publication data
Guillemin, V., 1937–

Symplectic fibrations and multiplicity diagrams / Victor
Guillemin, Eugene Lerman, Shlomo Sternberg.

p.   cm.

Includes bibliographical references and index.

ISBN 0-521-44323-7 (hc)

1. Group theory.   2. Quantum theory.   3. Representations of
groups.   4. Symplectic groups.   I. Lerman, Eugene.   II. Sternberg,
Shlomo.   III. Title.
QC174.17.G7G85   1996
530.1′2′0151255–dc20   95-45552
                                                             CIP

ISBN 978-0-521-44323-4 hardback
ISBN 978-0-521-11186-7 paperback

Contents

Acknowledgments

This book has an unofficial coauthor. Chapters 1, 2, and 3 were considerably revised and expanded after Yael Karshon went through the earlier versions in painstaking detail and suggested innumerable simplifications, clarifications, and stylistic improvements. The book in its present form is far from perfect, but without her assistance it would have been in far worse shape than it is! We would like to acknowledge here, as we will have frequent occasion to do elsewhere in the text, our debt to her.

Introduction

Multiplicity diagrams are convenient ways of depicting certain phenomena in the representation theory of Lie groups. Given a compact Lie group G and a unitary representation ρ of G on a finite-dimensional Hilbert space, ρ can be completely described by listing the multiplicity with which each irreducible representation of G occurs in ρ. Suppose now that we are given a scheme for indexing the irreducible representations of G. Then the multiplicity diagram attached to ρ will be a diagram that displays the indexing set and labels each point in the set with the multiplicity with which that point occurs in ρ. For instance, suppose $G = T^n =$ the real n-dimensional torus. Then the irreducible representations of G are indexed by the points on the n-dimensional lattice $\mathbb{Z}^n$, and a multiplicity diagram will consist of a subregion Δ of $\mathbb{Z}^n$-space with an integer-valued function m defined on it. The subregion Δ will be, essentially, the support of m. In the case of primary interest to us, the function m itself will be given by an explicit formula – the celebrated Kostant multiplicity formula. As is frequently the case with explicit formulas, the formula, although explicit, is extremely difficult to evaluate. The methods of this monograph will give us an effective way of making approximate computations with this formula using techniques from symplectic geometry. We will actually be interested less in the multiplicity diagrams themselves than in certain "asymptotic" versions of them. More specifically we will want to consider the following situation: a sequence of representations ρ_k, $k = 1, 2, \ldots$, such that as k tends to infinity the corresponding multiplicity diagrams Δ_k tend to a limit in some appropriate sense. Very often this will involve some rescaling of Δ_k so that the limit makes sense.

A simple example will make clearer what we mean by this: Let G be the n-torus and let ρ be the standard representation of G on the space of homogeneous polynomials of degree k in the variables $z^1, \ldots, z^n$. Under ρ the element

$g = (e^{i\phi_1}, \ldots, e^{i\phi_n})$ gets represented by the transformation

$$\sum a_N z^N \mapsto \sum a_N e^{i\phi \cdot N} z^N.$$

The irreducible representations associated with ρ are indexed by the multi-indices

$$N = (N_1, \ldots, N_n)$$

with $N_1, \ldots, N_n \geq 0$ and $N_1 + \cdots + N_n = k$. Moreover, each representation occurs with multiplicity one, so the multiplicity diagram for this representation consists of the integer points on the $(n-1)$-simplex

$$x_1 + \cdots + x_n = k, \quad x_1, \ldots, x_n \geq 0,$$

and m is the constant function one. If we rescale this diagram by multiplying each x by $1/k$ and let k tend to infinity we get as our "limit diagram" the standard $(n-1)$-simplex

$$x_1 + \cdots + x_n = 1, \quad x_1, \ldots, x_n \geq 0,$$

and m will be, as before, the constant function one.

In this monograph we will for the most part be concerned with examples that are only slightly more complicated than this simple example. In particular G will usually be the n-torus, and the ρs will usually be representations of the following sort: We will embed G in a larger group H, which is compact but not abelian, and take ρ to be the restriction to G of an irreducible representation of H. For these representations the corresponding multiplicity diagrams have some remarkable features: To begin with, the region Δ in $\mathbb{Z}^n$ on which m is supported turns out to be a convex polytope. (For instance, in the preceding example it is the standard $(n-1)$-simplex.) Moreover, one can decompose Δ into a union of convex subpolytopes

$$\Delta = \Delta_1 \cup \cdots \cup \Delta_n \tag{0.1}$$

such that m is equal to a polynomial on each Δ_i. This polynomial (which we will call the *interpolating polynomial* of m on the region Δ_i) has integer coefficients, and its degree is bounded by a number that depends on the dimension and rank of H and the dimension of G but is independent of ρ itself.

Thanks to some recent results of Duistermaat and Heckman quite a bit is known about these interpolating polynomials. In fact if we confine ourselves to asymptotic results of the kind already described, they can be more or less completely computed. We will give some recipes for computing them in Chapter 3 and work out quite a few explicit examples in Chapters 3, 4, and 5. One thing that we will see from these examples is that if G is the Cartan subgroup of H and ρ is a sufficiently "generic" representation of H, then for most of the Δ_is in the decomposition (0.1) the degree of the interpolating polynomial is exactly equal to $d = (\dim H - \operatorname{rank} H)/2$. For instance, it is exactly of degree d if Δ_i contains an exterior vertex (or side or face) in its closure. For a few of the Δ_is, however, this polynomial is of degree much less than d. We will call these regions (which will be, in fact, the regions that we are mainly interested in) *lacunary* regions. On these regions a lot of mysterious cancellations occur in the Duistermaat–Heckman formulas, and one of our main goals will be to explain why this happens and what role these regions play in the representation theory of H.

A large part of this monograph is concerned with questions in symplectic geometry which would seem at first glance to have little to do with the questions in representation theory just described, so we will say a few words about the reason for this: By the Bott–Borel–Weil theorem there is, roughly speaking, a one-to-one correspondence between the irreducible representations of H and certain coadjoint orbits of H. Moreover, by a theorem of Kirillov, Kostant, and Souriau the coadjoint orbits of H are, up to covering, exactly the symplectic manifolds on which H acts as a transitive group of symplectomorphisms. Now if ρ is an irreducible representation of H and $\mathcal{O}$ is the coadjoint orbit corresponding to it, the multiplicity function m is, it turns out, a symplectic invariant of $\mathcal{O}$. (This in fact is the main content of the Duistermaat–Heckman formulas. The details will be spelled out in Chapter 3.) Therefore, it is reasonable to suppose that lacunae are connected with phenomena in symplectic geometry, and indeed this is the case: It has long been known that there is a kind of "hierarchy" among the irreducible representations of a compact Lie group having to do with the symmetry properties of their Dynkin diagrams. At the level of coadjoint orbits this hierarchy is reflected in the existence of symplectic fibrations of certain coadjoint orbits over others. (We will describe this hierarchy in detail from this point of view in Chapter 2.) Our main result is that *it is these fibrations that are responsible for the existence of lacunae.* Namely, let ρ_1 and ρ_2 be irreducible representations of H and let $\mathcal{O}_1$ and $\mathcal{O}_2$ be the corresponding coadjoint orbits. If $\mathcal{O}_1$ fibers symplectically over $\mathcal{O}_2$ then the multiplicity diagram for ρ_1, albeit much more complicated than the multiplicity diagram for ρ_2, contains the multiplicity diagram for ρ_2 as a subdiagram (or at least contains a diagram very similar to it) and it is this diagram that looks like the "lacunae" in the diagram for ρ_1. The explanation for this will be given in the beginning of Chapter 4.

A few words about the content of this monograph: Chapter 1 consists of a detailed exposition of the theory of symplectic fibrations. Some of this material is known, but there seems to be no place in the literature where it is easily accessible. Chapter 2 discusses a lot of examples of such fibrations, in particular, of coadjoint orbits by coadjoint orbits. Chapter 3 contains an exposition of the Duistermaat–Heckman theory and, as an application of this theory, formulas for the interpolating polynomials. Chapter 4 has to do with lacunae and their connection with symplectic fibrations. (The main theorem in Chapter 4, of which the result on lacunae is a consequence, is that if a Lie group G acts in an appropriate way on a symplectic fibration, the Marsden–Weinstein reduction operation gives rise to a "reduced" symplectic fibration.) Finally, in Chapter 5 we discuss a number of examples having to do for the most part with groups of low rank. In particular we exhibit the multiplicity diagrams for the six-, eight-, and ten-dimensional orbits of $SU(4)$. (The diagrams for the twelve-dimensional orbits were unfortunately already too complicated for the graphics capabilities of our Macintoshes, but we will at least be able to give a rough idea of how they look.)

1

Symplectic Fibrations

1.1 What Is a Symplectic Fiber Bundle?

Our definition will be the following: Let F be a symplectic manifold and let

$$\pi\colon M \to B \tag{1.1}$$

be a differential fibration with standard fiber F. We will say that (1.1) is a symplectic fibration if there exists a covering $\{U_\alpha\}$ of B by open sets, and for each set U_α in the covering, a local trivialization

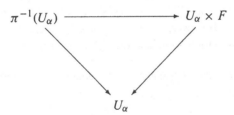

such that the transition maps

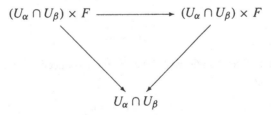

are symplectic mappings of $\{p\} \times F$ onto itself for every $p \in U_\alpha \cap U_\beta$. If F is compact one can replace this rather unwieldy definition by the following simpler one. For every $p \in B$ let $H^k(F, p)$ be the kth de Rham cohomology group of the fiber above p. The assignment $p \mapsto H^k(F, p)$ defines a smooth locally flat vector bundle $H^k(F)$ over B. (By "locally flat" we mean: equipped with a canonical locally flat connection. See, for instance, [Go].) Now suppose that ω_p is a symplectic form on the fiber above p that varies smoothly as one varies p.

Theorem 1.1.1 *A necessary and sufficient condition for $\pi\colon M \to B$ to be a symplectic fibration in the sense of the previous definition is that the section of the vector bundle $H^2(F)$ defined by $p \mapsto [\omega_p]$ be autoparallel.*

Proof. Let U be a contractible open subset of B and

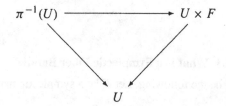

a local trivialization of M over U. Then the symplectic forms on the fibers above U define a family $\{\omega_b \mid b \in U\}$ of symplectic forms on F with the property that the cohomology class $[\omega_b]$ is the same for all b. The proof of the theorem comes down to showing that there exists a symplectic form ω on F and a diffeomorphism $\kappa_b\colon F \mapsto F$ depending smoothly on $b \in U$ such that $\kappa_b^*\omega = \omega_b$ for all b. However, since U is contractible and F compact, this can easily be proved by Moser's method. (See Theorem 1.6.2 to follow.)   □

In the next section we will make a slight change in our definition for non-compact fibrations.

1.2 Symplectic Connections

Let

$$\pi\colon M \to B \qquad\qquad (1.2)$$

be a symplectic fiber bundle with fiber F. By a connection on (1.2) we will mean a splitting

$$(\Gamma) \qquad\qquad TM = Vert \oplus Hor$$

into vertical and horizontal pieces. Given a curve γ on B through the point b_0, one gets from (Γ) local liftings of γ to horizontal curves in M. For any point $q \in F_{b_0}$, we thus get a holonomy map

$$\tau_\gamma\colon F_{b_0} \supset U_q \to F_b, \qquad\qquad (1.3)$$

which is defined in a neighborhood U_q of q in F_{b_0}, where $b \in B$ is a point further along the curve γ.

Definition 1.2.1 (provisional) *We will say that the connection* (Γ) *is* symplectic *if (1.3) is a symplectic mapping for all* $q \in F_{b_0}$ *and for all choices of* γ.

There is a very simple necessary and sufficient condition due to [GLSW] for symplecticity of (Γ), which we will discuss (and use to show that every symplectic fiber bundle admits such a connection). We will also show in Sections 3 and 4 that the existence of "nice" symplectic connections on M is closely related to the following question: "Does there exist a symplectic form on M whose restriction to each fiber is the preassigned symplectic form on that fiber?" Notice, by the way, that there always exists some two-form on M with this property. (Proof: Cover M by open sets $\{U_i\}$ such that on each U_i there exists an ω_i with this property. Let $\{\rho_i\}$ be a partition of unity subordinate to this cover. Then

$$\sum \rho_i \omega_i$$

is a globally defined two-form with this property.) Two-forms with this property will come up frequently in the discussion to follow, so it will be useful to give them a name.

Definition 1.2.2 *A two-form* ω *on* M *will be said to be* fiber-compatible *if its restriction to the fiber above each point of* B *is the given symplectic two-form on that fiber.*

If ω is fiber-compatible we can associate with it a connection on M by defining the horizontal bundle in (Γ) to be the orthogonal complement of the vertical bundle with respect to the bilinear form ω, that is, by defining the fiber of the horizontal bundle at p to be the space

$$Hor_p := \{v \in T_p M \mid \quad \omega(v, w) = 0 \text{ for all } w \in Vert_p\}. \tag{1.4}$$

Conversely it is clear that every connection on M can be defined this way. For example, given the connection (Γ), define the form ω by declaring $\omega(v, w) = 0$ whenever v is horizontal and for all w, while the restriction of ω to the fiber is taken to be the given symplectic form. Since the tangent space at any point of M decomposes into the direct sum of the horizontal and vertical subspaces, this completely defines ω and (1.4) recovers (Γ) as the connection associated to ω.

Definition 1.2.3 *If (1.4) is the horizontal space of the connection (Γ) at all $p \in M$, we will say that ω is (Γ)-compatible.*

Theorem 1.2.4 ([GLSW], theorem 4) *Let ω be a (Γ)-compatible two-form on M. Then the connection (Γ) is symplectic if and only if*

$$\iota(v_1 \wedge v_2)d\omega = 0 \qquad (1.5)$$

for every pair of vertical vector fields v_1 and v_2.

Proof. Let v be a vector field on B and let $v^\#$ be its horizontal lift to M. What we have to show is that locally on M, for small enough t,

$$(\exp tv^\#)^*\omega = \omega \quad (\mathrm{mod}\ B), \qquad (1.6)$$

where the "(mod B)" means that the restriction to each fiber of the right-hand side is equal to the restriction of the left-hand side. Differentiating (1.6) with respect to t reduces this to the condition

$$(L_{v^\#}\omega)(v_1, v_2) = 0 \qquad (1.7)$$

for every pair of vertical vector fields v_1 and v_2. However,

$$L_{v^\#}\omega = d\iota(v^\#)\omega + \iota(v^\#)d\omega \qquad (1.8)$$

by the Weil identity. We claim that the first term on the right of (1.8) contributes zero to (1.7). Indeed

$$(d\iota(v^\#)\omega)(v_1, v_2) = L_{v_1}(\iota(v^\#)\omega(v_2)) - L_{v_2}(\iota(v^\#)\omega(v_1)) - \iota(v^\#)\omega([v_1, v_2]),$$

which is zero since v_1, v_2, and $[v_1, v_2]$ are vertical and $v^\#$ is horizontal. On the other hand

$$(\iota(v^\#)d\omega)(v_1, v_2) = 0$$

for all v if and only if (1.5) holds. $\square$

We will show that there exists a two-form on M that is fiber-compatible and also satisfies the condition (1.5). In fact let $\{U_i\}$ be a covering of B by open sets with the property that the fibration is trivial over each U_i. It is clear that on the preimage of U_i there exists a two-form ω_i that is fiber-compatible and is closed. Let $\{\rho_i\}$ be a partition of unity subordinate to $\{U_i\}$. We leave it for the reader to check that

$$\sum \rho_i \omega_i$$

has both required properties – it is fiber-compatible and satisfies (1.5). Hence the connection defined by ω is symplectic. In particular this proves:

Theorem 1.2.5 *Every symplectic fiber bundle possesses a symplectic connection.*

We will now take Theorem 1.2.4 as our *definition* of a symplectic fibration:

Definition 1.2.6 *A fibration* $M \longrightarrow B$ *is a* symplectic fibration *if the fibers are all symplectic manifolds and if there exists a two-form ω on M that satisfies $\iota(v_1 \wedge v_2)d\omega = 0$ for every pair of vertical vector fields v_1 and v_2 and whose restriction to each fiber is the symplectic form of the fiber.*

As we have seen, each such ω gives rise to a connection on M, which is symplectic. If the fibers were compact, we could use this connection to transport the fiber F_{b_0} parallel over a point b_0 to fibers over nearby points and so get the local triviality conditions of section one. For noncompact fibers there are the usual problems of integrating the vector fields. But although Definition 1.2.3 is more natural, it is Definition 1.2.6 that we will work with in practice.

For example, if M is a symplectic manifold and $M \longrightarrow B$ is a fibration whose fibers are all symplectic submanifolds of M, then $M \longrightarrow B$ is a symplectic fibration.

1.3 Minimal Coupling

From now on we will assume that the connection (Γ) is symplectic. Let ω be a (Γ)-compatible two-form on M. In this section we will derive a formula for the curvature of (Γ) in terms of ω and $d\omega$. Before stating this formula, however, we will explain what we mean by the curvature of (Γ).

From the splitting

$$TM = Vert \oplus Hor$$

and the proof of Frobenius's theorem one gets a morphism of vector bundles

$$\kappa \colon \Lambda^2 Hor \to Vert \qquad (1.9)$$

which measures the extent to which the horizontal bundle fails to be integrable: If v_1 and v_2 are vector fields on B and $v_1^\#$ and $v_2^\#$ their horizontal lifts to M, then $\kappa(v_1^\#, v_2^\#)$ is the vertical component of $[v_1^\#, v_2^\#]$. Now let b be any point in B, and let T_b be the tangent space to B at b and F_b the fiber above b in M. Composing κ with the bijective map

$$Hor_p \cong T_b$$

for every $p \in F_b$, one gets a linear map

$$\Lambda^2 T_b \rightarrow \text{Vector Fields } (F_b). \qquad (1.10)$$

This mapping is, by definition, the *curvature* of (Γ) at b. Notice that since (Γ) is symplectic, the image of this map is contained in the Lie algebra of symplectic vector fields on F_b. Indeed, for each vector field v on B, the vector field $v^\#$ is fiberwise symplectic since the connection (Γ) is symplectic. Hence the vertical vector field

$$Vert[v_1^\#, v_2^\#] = [v_1^\#, v_2^\#] - [v_1, v_2]^\#$$

is symplectic on each fiber.

Now let ω be a (Γ)-compatible two-form on M and let v_1 and v_2 be vector fields on B. We will prove the following identity:

$$- d\iota(v_1^\#)\iota(v_2^\#)\omega + \iota(v_1^\#)\iota(v_2^\#)d\omega = \iota([v_1^\#, v_2^\#])\omega \quad (\text{mod } B). \qquad (1.11)$$

Here the "(mod B)" again means that the restriction to each fiber of the right-hand side is equal to the restriction of the left-hand side. Before we prove this identity, however, let us make a few observations about it. Notice first of all that the right-hand side is essentially the curvature of (Γ) (which justifies our calling (1.11) the *curvature identity*). Notice also that if ω is closed, then the identity reduces to

$$- d\iota(v_1^\#)\iota(v_2^\#)\omega = \iota([v_1^\#, v_2^\#])\omega \quad (\text{mod } B). \qquad (1.12)$$

This identity says that on the fiber F_b over a point b in B the vertical part of $[v_1^\#, v_2^\#]$ is a globally Hamiltonian vector field, its Hamiltonian function being the restriction of $-\omega(v_1^\#, v_2^\#)$ to F_b. This remarkable relationship between the curvature of Γ and the horizontal part of the two-form is known as *minimal coupling*. (See [S1].)

Let us now prove (1.11). Let v be a vertical vector field on M. Then

$$L_v\big[\omega(v_1^\#, v_2^\#)\big] = (L_v\omega)(v_1^\#, v_2^\#) + \omega\big([v, v_1^\#], v_2^\#\big)$$
$$+ \omega\big(v_1^\#, [v, v_2^\#]\big)$$
$$= [\iota(v)d\omega + d\iota(v)\omega](v_1^\#, v_2^\#),$$

since $[v, v_1^\#]$ and $[v, v_2^\#]$ are vertical. For any one-form θ and any vector fields w_1 and w_2, the Weil formula implies that

$$d\theta(w_1, w_2) = \iota(w_1)d[\theta(w_2)] - \iota(w_2)d[\theta(w_1)] - \theta([w_1, w_2]).$$

For $\theta = \iota(v)\omega$, $w_1 = v_1^\#$, $w_2 = v_2^\#$ the first two terms vanish, so

$$\iota(v)d\big[\omega(v_1^\#, v_2^\#)\big] = L_v\big[\omega(v_1^\#, v_2^\#)\big] = d\omega(v_1^\#, v_2^\#, v) - \omega\big(v, [v_1^\#, v_2^\#]\big),$$

giving (1.11).   □

1.4 The Coupling Form ω_Γ

Let us now return to the question that we posed in Section 1.2. If Γ is a symplectic connection, is it possible to find a closed Γ-compatible two-form on M? We will prove that if F is compact and simply connected there exists a canonical such two-form. First of all, however, we will make a few comments about these hypotheses. Let F be a symplectic manifold with symplectic form ω_F. Each smooth function ϕ on F determines a unique vector field v_ϕ by

$$\iota(v_\phi)\omega_F = -d\phi.$$

(Two functions that differ by a constant determine the same vector field.) This then defines a Poisson bracket on the space of functions by

$$\{\phi, \psi\} := v_\phi \psi = \iota(v_\phi)d\psi = -\iota(v_\phi)\iota(v_\psi)\omega_F.$$

Then

$$\begin{aligned}
\iota\big([v_\phi, v_\psi]\big)\omega_F &= L_{v_\phi}\iota(v_\psi)\omega_F \\
&= -L_{v_\phi}d\psi \\
&= -dv_\phi\psi \\
&= -d\{\phi, \psi\},
\end{aligned}$$

so

$$v_{\{\phi,\psi\}} = \big[v_\phi, v_\psi\big];$$

the map $\phi \mapsto v_\psi$ is a homomorphism from the algebra of smooth functions under Poisson bracket to the algebra of symplectic vector fields under Lie bracket. A vector field is symplectic if and only if $L_v\omega_F = 0$. Since $L_v\omega_F = d\iota(v)\omega_F + \iota(v)d\omega_F = d\iota(v)\omega_F$, we see that a vector field v is symplectic if and only if $\iota(v)\omega_F$ is closed. Hence, if $H^1(F, \mathbb{R}) = 0$, the homomorphism $\phi \mapsto v_\phi$ is surjective. If, in addition, F is compact, this map has a canonical inverse: For each symplectic vector field v there exists a unique C^∞ function ϕ^v with the properties

$$\iota(v)\omega_F = -d\phi^v$$

and

$$\int \phi^v \omega_F^d = 0,$$

d being half the dimension of F. This inverse, which we will denote by Φ_F, determines the *moment map* associated with the group of symplectomorphisms of F: each $p \in F$ gives a linear function l_p on the Lie algebra of symplectic vector fields according to the rule

$$l_p(v) = \Phi_F(v)(p).$$

Our main result is the following:

Theorem 1.4.1 *If F is compact, connected, and simply connected, there exists a unique closed Γ-compatible two-form ω on M with the property*

$$\pi_* \omega^{d+1} = 0, \tag{1.13}$$

π_* *being the Gysin or "fiber integration" map and d being half the dimension of the fiber.*

(For a $(2d + r)$-form α on M, $\pi_* \alpha$ is the r-form on B given by

$$\iota(v_1 \wedge \cdots \wedge v_r)\pi_*\alpha|_b = \int_{F_b} \iota(\tilde{v}_1 \wedge \cdots \wedge \tilde{v}_r)\alpha,$$

where $\pi_*(\tilde{v}_j) = v_j$.)

Proof. 1. (Uniqueness) Suppose ω exists. Let v_1 and v_2 be vector fields on B. By (1.12)

$$-d\iota\big(v_1^\#\big)\iota\big(v_2^\#\big)\omega = \iota\big([v_1^\#, v_2^\#]\big)\omega \quad (\text{mod } B).$$

Let b be a point in B and F_b the fiber above b in M. Equation (1.12) says that the image of $v_1(b) \wedge v_2(b)$ with respect to the curvature mapping (1.10) is a globally Hamiltonian vector field, with Hamiltonian function the restriction of $\omega(v_1^\#, v_2^\#)$ to F_b. Therefore, $\omega(v_1^\#, v_2^\#)$ is determined by Γ up to an additive constant (which depends on b). The Gysin condition asserts that, for all b,

$$\int_{F_b} i_b^*\big(\omega(v_1^\#, v_2^\#)\omega^d\big) = 0$$

(i_b^* being the restriction of the $2d$-form in parentheses to F_b) and hence clearly fixes this constant. This shows that the horizontal component of ω is completely determined by the curvature identity and the Gysin condition. However, the fact that ω is Γ-compatible implies that its verti-zontal component is zero, and its vertical component, of course, has to be ω_F. Thus ω is entirely determined by the curvature identity and the Gysin condition.

2. (Existence) As we just pointed out, the vertical and verti-zontal components of ω are specified by the condition of Γ-compatibility; so we only have to define the horizontal component of ω. Given vector fields v_1 and v_2 on B, we will define $-\omega(v_1^\#, v_2^\#)$ by defining its restriction to F_b to be the image of $v_1(b) \wedge v_2(b)$ with respect to the composite mapping

$$\Lambda^2(T_b) \to \Sigma(F_b) \to C^\infty(F_b),$$

the first arrow being the curvature map (1.10) and the second arrow Φ_{F_b}. If we denote by Φ the map that is equal to Φ_{F_b} on the fiber above b, we can write the definition of $\omega(v_1^\#, v_2^\#)$ more succinctly in the form

$$- \omega\big(v_1^\#, v_2^\#\big) = \Phi(Vert\,[v_1^\#, v_2^\#]). \tag{1.14}$$

We will now prove that ω is closed, that is, that $d\omega = 0$. From the results of Section 1.2 we know already (Theorem 1.2.4) that

$$\iota(w_1 \wedge w_2)d\omega = 0$$

for every pair of vertical vector fields w_1 and w_2. We have defined ω so that (1.12) holds. Hence, by the curvature identity (1.11) it then follows that

$$\iota(w)d\omega = 0$$

for every vertical vector field w. Therefore, to show that $d\omega = 0$ it suffices to show that

$$d\omega(v_1^\#, v_2^\#, v_3^\#) = 0$$

for every triple of vector fields $v_1, v_2,$ and v_3 on B. To prove this we will again use the standard formula for $d\omega(v_1^\#, v_2^\#, v_3^\#)$:

$$d\omega(v_1^\#, v_2^\#, v_3^\#) = -\omega\big([v_1^\#, v_2^\#], v_3^\#\big) - \cdots + L_{v_1^\#}\omega(v_2^\#, v_3^\#) + \cdots, \tag{1.15}$$

the dots indicating cyclic permutations. Notice that

$$[v_1^\#, v_2^\#] = Vert\,[v_1^\#, v_2^\#] + [v_1, v_2]^\#. \tag{1.16}$$

Thus $-\omega([v_1^\#, v_2^\#], v_3^\#)$ is equal to

$$-\omega\big(Vert\,[v_1^\#, v_2^\#], v_3^\#\big) - \omega([v_1, v_2]^\#, v_3^\#)$$

and the first of these terms is zero since the verti-zontal part of ω is zero. Now let's use (1.14) to replace the second term by

$$\Phi\big(Vert\,[[v_1, v_2]^\#, v_3^\#]\big) \tag{1.17}$$

and again make the substitution (1.16) and use the fact that the bracket of a field of the form $v^{\#}$ and a vertical vector field is vertical. This allows us to replace (1.17) by

$$\Phi\left(Vert\left[[v_1^{\#}, v_2^{\#}], v_3^{\#}\right]\right) + \Phi\left(L_{v_3^{\#}} Vert\left[v_1^{\#}, v_2^{\#}\right]\right). \qquad (1.18)$$

Finally notice that with the derivation moved outside, the second term in (1.18) becomes

$$-L_{v_3^{\#}}\omega(v_1^{\#}, v_2^{\#}).$$

Summarizing, we have shown that

$$\omega\left([v_1^{\#}, v_2^{\#}], v_3^{\#}\right) - L_{v_3^{\#}}\omega(v_1^{\#}, v_2^{\#}) = -\Phi\left(Vert\left[[v_1^{\#}, v_2^{\#}], v_3^{\#}\right]\right).$$

The term on the right plus its cyclic permutations vanishes by the Jacobi identity. This shows that (1.15) is zero and hence that $d\omega = 0$. $\qquad\square$

Remark. The hypothesis that F be compact and simply connected can be replaced by much weaker hypotheses. Suppose, for instance, that the holonomy group G of the connection Γ is a finite-dimensional Lie group. Then, by the Ambrose–Singer theorem (cf. [AS]), a necessary and sufficient condition for there to exist a closed Γ-compatible two-form is that the action of G on F be Hamiltonian, that is, that there exist a G-equivariant Lie algebra homomorphism $\Psi\colon \mathfrak{g} \to C^{\infty}(F)$ making the following diagram commute:

$$C^{\infty}(F) \;\to\; \Sigma(F)$$
$$\nwarrow \quad \uparrow$$
$$\mathfrak{g}$$

In fact one can prove the *sufficiency* of this condition by substituting this slanted arrow everywhere that Φ_F occurs in the previous proof. Of course the condition (1.13) is then replaced by condition (1.14), with Ψ instead of Φ, and now ω depends on the choice of Ψ giving the Hamiltonian action.

An important example of a Hamiltonian action is the following: Let Y be a differentiable manifold and $F = T^*Y$. If $G = \mathrm{Diff}\, Y$ then G acts on T^*Y: Each $\psi \in \mathrm{Diff}\, Y$ has a differential

$$d\psi_y\colon TY_y \;\longrightarrow\; TY_{\psi(y)}.$$

The dual map is

$$d\psi_y^*\colon T^*Y_{\psi(y)} \;\longrightarrow\; T^*Y_y.$$

We let $\psi^*: T^*Y \longrightarrow T^*Y$ be the diffeomorphism whose restriction to $T^*Y_{\psi(y)}$ is $d\psi^*_y$. Then

$$\psi \rightsquigarrow (\psi^*)^{-1}$$

defines an action of Diff Y on T^*Y. If α denotes the canonical one-form on T^*Y then this action of Diff Y on T^*Y preserves α. The symplectic form on T^*Y is $\omega = d\alpha$. The Lie algebra of Diff Y is $V(Y)$, the algebra of all vector fields on Y. Each vector field ξ on Y determines a function

$$f_\xi(z) = -\langle z, \xi_{\pi(z)} \rangle$$

on T^*Y and this defines a Hamiltonian action of Diff Y on T^*Y.

So if we have a symplectic fibration $M \longrightarrow B$ whose typical fiber F is a cotangent bundle, $F = T^*Y$, and if the connection Γ has the property that its holonomy group is a subgroup of Diff Y, then we get an ω. At the beginning of Chapter 2 we will show that if $X \longrightarrow B$ is a differentiable fibration (with ordinary manifolds X_b as fibers) then a connection A on X gives rise to connection Γ on the symplectic fibration whose fibers are T^*X_b. Furthermore, Γ will have the properties described earlier, and hence give rise to our ω.

Definition 1.4.2 *Henceforth, to indicate the dependence of the form ω obtained in Theorem 1.4.1 on the connection Γ, we will denote it by ω_Γ, and we will call it the* coupling form *associated with the symplectic connection Γ.*

The form ω_Γ is not the only closed Γ-compatible two-form. In fact if ω_B is a closed two-form on B then

$$\omega_\Gamma + \pi^*\omega_B \tag{1.19}$$

is another such form. We will now prove a converse of this result which has a number of interesting implications.

Theorem 1.4.3 *Every closed Γ-compatible two-form on M is of the form (1.19).*

Proof. Let ω' be such a form. Then, for every vertical vector field v,

$$\iota(v)(\omega' - \omega_\Gamma) = \iota(v)d(\omega' - \omega_\Gamma) = 0.$$

Hence

$$L_v(\omega' - \omega_\Gamma) = 0.$$

This equation and the first of the two preceding equations can be interpreted as saying that $\omega' - \omega_\Gamma$ is *basic*, that is, that there exists a closed two-form ω_B

on B such that

$$\omega' - \omega_\Gamma = \pi^* \omega_B.$$

In other words ω' is of the form (1.19). □

We will now discuss some implications of the result. One is that there exists no nontrivial symplectic analog of a notion that plays a fundamental role in Riemannian geometry, that of a *Riemannian submersion*: Given a fiber bundle $\pi \colon M \to B$ and a Riemannian metric on M, the metric defines an orthogonal splitting

$$TM = \text{Vert} \oplus \text{Hor}.$$

The pair (M, π) is called a Riemannian submersion if there exists a metric g_B on B such that the horizontal part of the metric on M is $\pi^* g_B$. Suppose now that the symplectic form (1.19) had the property that its horizontal part was ω_B, that is, that the horizontal part of ω_Γ was zero. By the curvature identity (1.12) this would say that the curvature of the connection Γ is zero. Hence, if B were simply connected, M would have to be the symplectic product of B and F.

An even more important consequence is the following. Suppose ω is a closed two-form on M whose restriction to the fiber F_p over every point $p \in B$ is symplectic. Then, by Definition 1.2.6, (M, B, π) is a symplectic fibration and, by Theorem 1.2.4, ω defines a symplectic connection Γ on (M, B, π). Moreover, by (1.19) there exists a presymplectic (i.e., a closed but not necessarily nondegenerate) form ω_B on B such that $\omega = \omega_\Gamma + \omega_B$. Thus one can think of (1.19) as a recipe for converting presymplectic forms ω on M into presymplectic forms ω_B on the base B, in other words *as a kind of symplectic analog of the operation of reduction!* (For more on the latter see Section 3.1.)

1.5 Weak Coupling

The coupling form ω_Γ will not necessarily be symplectic even though its restriction to each fiber is symplectic (the problem being that it may degenerate in horizontal directions). At first glance it might appear that the way to get around this problem is to add a horizontal correction term to ω_Γ as in (1.19). However, since ω_Γ may have a nonzero horizontal component itself, it's still not clear that the resulting form will be symplectic, even if the ω_B in (1.19) is symplectic. If M is compact, however, one way to get a symplectic structure on M is via *weak coupling*: Rescale the symplectic form on each fiber by multiplying it by ϵ, ϵ being a fixed (but small) positive constant. The effect of this rescaling on the coupling form is to multiply it by ϵ as well; so (1.19) gets replaced by

$$\omega_\epsilon = \epsilon \omega_\Gamma + \pi^* \omega_B \qquad (1.20)$$

and it's clear that (1.20) is symplectic if ω_B is symplectic on B and ϵ is small enough.

Definition 1.5.1 *The form (1.20) is the* weak coupling form *associated with the symplectic connection Γ and the symplectic form ω_B on B.*

Weak coupling has to do with the behavior of objects on M associated with the symplectic structure (1.20) as ϵ tends to zero. In Chapter 4 we will discuss (in considerable detail) weak coupling in connection with the computation of Duistermaat–Heckman invariants. We will now say a few words about weak coupling in connection with Hamiltonian systems. For example, let H be a real-valued smooth function on B. The symplectic structure on B then associates to H a Hamiltonian vector field, call it v_B. We can consider the function $H \circ \pi$ on M, and associate to it a Hamiltonian vector field v_ϵ relative to the symplectic form ω_ϵ. In physics v_ϵ describes the dynamics of a classical system in the presence of a "gauge field" represented by the connection Γ. Weak coupling describes the effect of the gauge field on the system when the gauge forces are small in comparison with the other forces acting on the system. Notice that the equation

$$\iota(v_\epsilon)\omega_\epsilon = -d\,\pi^* H \qquad (1.21)$$

implies that v_ϵ is horizontal. It also follows from (1.20) and (1.21) that if we set

$$v_0 = \lim_{\epsilon \to 0} v_\epsilon$$

then v_0 is the horizontal lift of v_B:

$$v_0 = v_B^{\#}.$$

In particular, its integral curves are the horizontal lifts of the integral curves of v_B. For example, suppose that γ is a periodic integral curve of v_B with $\gamma(0) = b$. Let $\tau_\gamma \colon F \longrightarrow F$ be the holonomy element associated with γ. If p is a point in M sitting above b then the horizontal lift of γ through p will close up when γ closes up if and only if $\tau_\gamma(p) = p$, that is, if and only if p is a fixed point of τ_γ. Thus there is a one-to-one correspondence between fixed points of τ_γ and periodic integral curves of v_0 sitting over γ.

Suppose now that γ lies on the constant-energy surface $H = E$ and is a *nondegenerate* periodic trajectory of v_B, that is, there are no "infinitesimally nearby" periodic trajectories on the energy surface $H = E$. (In technical language this means the following. Let T_b^E be the tangent space to the energy surface at b and let

$$P_\gamma : T_b^E \longrightarrow T_b^E$$

be the linearized Poincaré map associated with the flow near γ. (See [AM].) The vector $v_B(p)$ is an eigenvector of P_γ with eigenvalue 1. If this vector is the *only* eigenvector with eigenvalue one, in other words, if $P_\gamma - \mathrm{Id}$ is of corank one, then γ is said to be nondegenerate.) Suppose in addition that p is a nondegenerate fixed point of τ_γ. (Here we mean nondegenerate in the usual "Lefschetz" sense: $(d\tau_\gamma)_p - \mathrm{Id}$ is bijective.) Then γ_0 is a nondegenerate periodic integral curve of the vector field v_0 on the surface $H \circ \pi = E$, and by standard perturbation techniques it follows that, for all ϵ sufficiently small, there exists a unique nondegenerate periodic integral curve γ_ϵ of the vector field v_ϵ on the energy surface $H \circ \pi = E$ such that γ_ϵ depends smoothly on ϵ and such that $\gamma_\epsilon = \gamma_0$ when $\epsilon = 0$. In particular, if the fixed points of τ_γ are all nondegenerate and are K in number, this construction gives rise to K periodic trajectories of v_ϵ. These trajectories will, of course, be far apart in M but their projections onto B will be very close to the original trajectory γ (so close in fact that, if the external field is very small, one will need very sensitive detection devices to tell them apart from the original trajectory).

Example 1.1 In what physicists call GUTs or grand unified theories, B is the cotangent bundle of Minkowski space, and the data that are used to construct M are a principal $SU(5)$-bundle

$$P \longrightarrow B$$

equipped with an $SU(5)$-invariant connection.

In the next chapter we will show how to construct from this data and from a coadjoint orbit $\mathcal{O}$ of $SU(5)$ a symplectic fibration $M \longrightarrow B$ with fiber $\mathcal{O}$ and a symplectic connection Γ. In this example, $\tau_\gamma : \mathcal{O} \longrightarrow \mathcal{O}$ will turn out to be the action of an element g_γ of $SU(5)$ on $\mathcal{O}$. If g_γ is a regular element of $SU(5)$ and $\mathcal{O}$ an orbit of maximal dimension, the action of g_γ on $\mathcal{O}$ will have nondegenerate fixed points, and they will be 120 in number. Thus in the weak coupling limit, GUT theories predict the existence of 120 periodic classical trajectories in the vicinity of every classical periodic trajectory of the "ungauged" theory.

One can consider weak coupling in conjunction with "scaling and the adiabatic limit" as follows: Suppose that the fibration $M \to B$ has an $\mathbb{R}^+$ action that preserves all fibers and that multiplies the symplectic form on each fiber by a scale factor. For example, if B is a point and $F = T^*Y$ then the transformation

$$m_\lambda \colon T^*Y \longrightarrow T^*Y, \qquad m_\lambda(y, \eta) = (y, \lambda\eta)$$

has the property that

$$m_\lambda^* \omega = \lambda\omega,$$

where ω is the canonical symplectic form on T^*Y. More generally, we will have such an action with the same scaling properties whenever the fibers are cotangent bundles in a consistent way, for example, if $X \to B$ is a fibration with an ordinary manifold X_b as fiber over each $b \in B$ and we let M be the fibration whose fiber over each b is T^*X_b. If the action of $\mathbb{R}^+$ preserves the connection Γ, it is reasonable to expect that the coupling form ω_Γ also scales as

$$m_\lambda^* \omega_\Gamma = \lambda\omega_\Gamma. \tag{1.22}$$

In the next chapter we shall exhibit an important collection of examples where (1.22) holds in conjunction with Riemannian submersions with totally geodesic fibers. For the moment let us take (1.22) as a hypothesis. Since $\mathbb{R}^+$ acts trivially on B, (1.22) implies that

$$m_\epsilon^* (\omega_\Gamma + \pi^* \omega_B) = \epsilon\omega_\Gamma + \pi^* \omega_B. \tag{1.23}$$

In other words, weak coupling can be seen as the effect of scaling, that is, the action of $\mathbb{R}^+$. Let us now go back to equation (1.21) with a slight change of notation, in that we will denote the function H occurring on the right-hand side by H_B to emphasize that it comes from the base, in particular,

$$m_\epsilon^* H_B = H_B. \tag{1.24}$$

Putting together (1.21), (1.23), and (1.24) we get

$$\iota(v_\epsilon) m_\epsilon^* (\omega_\Gamma + \pi^* \omega_B) = -dH_B = -m_\epsilon^* dH_B$$

or

$$m_\epsilon^* [\iota(m_{\epsilon-1}^* v_\epsilon)(\omega_\Gamma + \pi^* \omega_B)] = -m_\epsilon^* dH_B.$$

Applying $m_{\epsilon-1}^*$ to both sides we get $m_{\epsilon-1}^* v_\epsilon = v_1$ or

$$v_\epsilon = m_{\epsilon*} v_1.$$

Now let us replace the hypothesis that H is invariant under scaling by the following assumption: Suppose that

$$H = H_B + H_F,$$

where H_B satisfies (1.24) as before and where H_F satisfies

$$m_\epsilon^* H_F = \epsilon^2 H_F \tag{1.25}$$

and

$$\iota(v^\#) d H_F = 0 \quad \text{for all horizontal vector fields} \quad v^\#. \tag{1.26}$$

As we shall see in Chapter 2, these hypotheses are satisfied when H is the Hamiltonian associated to the Riemannian metric on a manifold X fibered over a manifold N with fibers that are totally geodesic. In this case $M = T^* X$, H_F is associated to the metric along the fibers, and H_B is associated to the metric along the base. We have the vector fields v_1 and w_1 defined by

$$\iota(v_1)(\omega_\Gamma + \pi^* \omega_B) = -d H_B$$

and

$$\iota(w_1)(\omega_\Gamma + \pi^* \omega_B) = -d H_F,$$

so the vector field associated to the Hamiltonian H and the symplectic form $\omega = \omega_\Gamma + \pi^* \omega_B$ is $v_1 + w_1$.

Proposition 1.5.2 *The vector field w_1 is vertical; in fact, on each fiber it is the vector field associated to the restriction of H to the fiber by the symplectic form of the fiber. In particular, $\{H_F, H_B\} = 0$ so the vector fields v_1 and w_1 commute.*

Proof. Since the horizontal vectors are the orthogonal complements of the vertical vectors with respect to ω, and since ω is symplectic, to prove the first assertion we must show that

$$\omega(v^\#, w_1) = 0$$

for all horizontal vector fields $v^\#$. But since $\iota(w_1)\omega = -d H_F$ this is precisely the content of assumption (1.26). The remaining assertions follow immediately. $\qquad\square$

Proposition 1.5.2 gives us a very good picture of the flow associated to H by ω. Namely, it is the superposition of two commuting flows, the minimal coupling flow generated by the vector field v_1, associated to the Hamiltonian H_B, and the vertical flow generated by w_1. The projection onto B of the trajectories associated to H is the same as the projection onto B of the trajectories associated to H_B. If we think of the fibers as representing "internal variables" that affect the motion of particles on B, then the trajectories on B relative to H_B and ω_B are modified by the presence of ω_Γ to the projections of solutions of the vector field v_1. Even though the flow generated by w_1 can be quite complicated, an observer on B does not see the effect of w_1.

Now let us consider a rescaled Hamiltonian H_ϵ defined by

$$H_\epsilon = H_B + \epsilon^{-2} H_F. \tag{1.27}$$

(In the case of a Riemannian submersion, this corresponds to multiplying the lengths in the fiber direction by a factor of ϵ, so that, as $\epsilon \to 0$, the lengths in the base direction become large relative to lengths in the fiber direction. This is known as the adiabatic limit; cf. the discussion in Chapter 2.) The vector field Ξ_{H_ϵ} corresponding to H_ϵ (relative to the minimal coupling form ω) is given by

$$\Xi_{H_\epsilon} = v_1 + \epsilon^{-2} w_1.$$

On the other hand, we can consider the vector field ζ_ϵ associated with the Hamiltonian H, but relative to the weak coupling form $\epsilon \omega_\Gamma + \pi^* \omega_B$, so

$$\iota(\zeta_\epsilon)(\epsilon \omega_\Gamma + \pi^* \omega_B) = -dH. \tag{1.28}$$

We claim that these two vector fields are related by

$$m_{\epsilon *}(\Xi_{H_\epsilon}) = \zeta_\epsilon. \tag{1.29}$$

Indeed, the argument is the same as before:

$$m_\epsilon^* [\iota(m_{\epsilon^{-1} *} \zeta_\epsilon)(\omega_\Gamma + \pi^* \omega_B)] = -dH = -m_\epsilon^* m_{\epsilon^{-1}}^* dH = -m_\epsilon^* dH_\epsilon,$$

proving (1.29). Equation (1.29) gives the relation between the adiabatic limit and the weak coupling limit.

1.6 Varying the Connection

Let us now address the following issue: How does the symplectic structure on M defined by (1.20) depend on the choice of the connection Γ? For simplicity we will, henceforth, assume that F is compact and simply connected.

Theorem 1.6.1 *Let Γ_1 and Γ_2 be symplectic connections on M and ω_1 and ω_2 the coupling forms associated with them. Then there exists a one-form μ on M with the properties*

$$\iota(v)\mu = 0$$

for every vertical vector field v and

$$\omega_2 - \omega_1 = d\mu.$$

Proof. Let $\kappa = \omega_2 - \omega_1$, and fix a coordinate patch U on B with coordinates $x_1, \ldots, x_n$. Then on the preimage of U

$$\kappa = \alpha_1 \wedge dx_1 + \cdots + \alpha_n \wedge dx_n$$

and since $d\kappa = 0$, $d\alpha_i = 0 \pmod{B}$. Therefore, since F is simply connected, there exist functions $f_1, \ldots, f_n$ such that $\alpha_i = df_i \pmod{B}$ (i.e., there exist one-forms $\beta_1, \ldots, \beta_n$ such that

$$\alpha_i = df_i + \beta_i$$

and $\langle v, \beta_i \rangle = 0$ for all vertical vector fields v). Substituting these expressions into the foregoing equation, we get for κ the identity

$$\kappa = \sum df_i \wedge dx_i + \sum \beta_i \wedge dx_i.$$

Since κ is closed and the first term in this sum is exact, the second term is also closed, and hence depends only on the x_is (not on the fiber variables). However, U is contractible so we can write this term in the form $d(\sum h_i dx_i)$, the h_is being functions of the x_is alone. Replacing f_i by $f_i + h_i$ we get for κ the formula

$$\kappa = d\left(\sum f_i dx_i \right).$$

Let $\mu_U = \sum f_i dx_i$. Then $\kappa = d\mu_U$ and $\iota(v)\mu_U = 0$ for all vertical vector fields v. Let U be a covering of B by such Us and $\{\rho_U\}$ a partition of unity subordinate to this cover. If we set

$$\mu = \sum \rho_U \mu_U$$

then

$$
\begin{aligned}
d\mu &= \sum \rho_U \, d\mu_U + \sum d\rho_U \wedge \mu_U \\
&= \sum \rho_U \kappa + \sum d\rho_U \wedge \mu_U \\
&= \kappa + \sum d\rho_U \wedge \mu_U.
\end{aligned}
$$

Thus $\iota(v)\mu = 0$, and $\iota(v)(\kappa - d\mu) = 0$ for all vertical vector fields v. Let $v = \kappa - d\mu$. Then, since v is closed and $\iota(v)v = 0$ for all vertical vector fields v, v is *basic*: There exists a closed two-form β on B such that $v = \pi^*\beta$. We will prove that the cohomology class $[\beta]$ is zero. To see this, let's rewrite κ as $\omega_2 - \omega_1$ and rewrite our equation as

$$\omega_2 = \omega_1 + d\mu + \pi^*\beta.$$

Then

$$\pi_*[\omega_2^{d+1}] = \pi_*[\omega_1^{d+1}] + (d+1)\pi_*[\omega_1^d \wedge \pi^*\beta]$$
$$= \pi_*[\omega_1^{d+1}] + (d+1)[\beta]\pi_*[\omega_1^d].$$

However, the image under the Gysin map of $[\omega_1^d]$ is a nonzero constant, and the images of $[\omega_2^{d+1}]$ and $[\omega_1^{d+1}]$ are zero since ω_2 and ω_1 are coupling forms; so $[\beta]$ is zero as claimed.

Thus there exists a one-form α on B such that $\beta = d\alpha$; and replacing μ by $\mu + \pi^*\alpha$ we obtain $\kappa = d\mu$, proving Theorem 1.6.1. $\qquad\square$

Now let ω_B be a symplectic form on B. Then, for ϵ sufficiently small,

$$\epsilon(\omega_1 + t\, d\mu) + \pi^*\omega_B$$

is symplectic for all t on the interval $[0,1]$. We recall now the following result of Moser (See, for example, [GS8]).

Theorem 1.6.2 *Let M be a compact manifold and let Ω_1 and Ω_0 be two symplectic forms on M. Suppose:*

i. $t\Omega_1 + (1-t)\Omega_0$ is symplectic for all $t \in [0,1]$.
ii. The cohomology classes $[\Omega_1]$ and $[\Omega_0]$ are equal.

Then Ω_1 and Ω_0 are symplectomorphic.

Proof. Let $\Omega_t = t\Omega_1 + (1-t)\Omega_0$. We will look for a family of diffeomorphisms $\{\Phi_t : M \to M\}_{t \in [0,1]}$ such that $\Phi_0 = \mathrm{Id}$ and

$$\Phi_t^*\Omega_t = \Omega_0 \qquad (1.30)$$

for all $t \in [0,1]$.

Suppose such a family exists. Then

$$0 = \frac{d}{dt}\Phi_t^*\Omega_t = \Phi_t^*\left(L_{\Xi_t}\Omega_t + \frac{d}{dt}\Omega_t\right)$$
$$= \Phi_t^*(d\iota(\Xi_t)\Omega_t + \Omega_1 - \Omega_0)$$

where Ξ_t is the time-dependent vector field generating $\{\Phi_t\}$. Since $[\Omega_1] = [\Omega_0]$, the there exists a one-form α such that $d\alpha = \Omega_1 - \Omega_0$. Thus if the isotopy $\{\Phi_t : M \to M\}_{t \in [0,1]}$ with the required property does exist, it must satisfy

$$d\alpha = -d\iota(\Xi_t)\Omega_t. \qquad (1.31)$$

On the other hand, since Ω_t is symplectic for all t, we can *define* a time-dependent vector field Ξ_t on M such that

$$\alpha = -\iota(\Xi_t)\Omega_t$$

for all $t \in [0, 1]$. Then (1.31) holds, and the corresponding isotopy $\{\Phi_t\}$ satisfies (1.30). In particular

$$\Phi_1^*\Omega_1 = \Omega_0,$$

providing us with the desired symplectomorphism.   □

Combining this result with Theorem 1.6.1 we get as a corollary:

Theorem 1.6.3 *Assume that M is compact. Then for all sufficiently small ϵ the symplectic forms $\epsilon\omega_2 + \pi^*\omega_B$ and $\epsilon\omega_1 + \pi^*\omega_B$ are symplectomorphic. In other words, in the weak coupling limit there is a unique symplectic structure on M compatible with the symplectic structure on fiber and base.*

2

Examples of Symplectic Fibrations:
The Coadjoint Orbit Hierarchy

As we pointed out in the introduction the purpose of this monograph is to explore connections between two subjects which seem to have, on the face of it, little to do with each other: symplectic fibrations and the multiplicity diagrams associated with representations of Lie groups. What will supply the bridge between these two topics is the coadjoint orbit hierarchy. There are several ways of describing this hierarchy, but the one we will give here is probably the simplest: A coadjoint orbit $\mathcal{O}_1$ will sit higher in this hierarchy than $\mathcal{O}_2$ if there exists an equivariant symplectic fibration of $\mathcal{O}_1$ over $\mathcal{O}_2$.[1] Thus, at the bottom of the hierarchy will sit the coadjoint orbits that can't be fibered over lower-dimensional coadjoint orbits. For example, for $SU(n)$ these are the complex Grassmannians and for $SO(2n)$ they include the complex quadrics and the "isotropic" Grassmannian. These minimal orbits correspond in representation theory to the fundamental representations of the groups $SU(n)$ and $SO(2n)$ and will play the same role here as their representation-theoretic counterparts.

To get a feeling for this hierarchy, it will be helpful to take a look at other examples of symplectic fibrations as well, and, in fact, without wandering too far from the goal we have in mind – the attempt to explain "lacunae" in multiplicity diagrams – we will discuss quite a few examples of symplectic fibrations that have little to do with coadjoint orbits and representation theory. (These include, in particular, two examples about which quite a bit is being written at the moment: Riemannian fibrations with totally geodesic fibers and symplectic fibrations with the Hannay–Berry connection. We haven't attempted to give a definitive description of either of these important examples but we have indicated how they are special cases of the general constructions outlined in the next few pages.)

---

[1] We exclude the zero orbit from the consideration.

What follows is a list of nine examples of symplectic fibrations. This list itself is a bit hierarchical, at the top being the most simple-minded examples and at the bottom the two rather sophisticated examples we just mentioned. Those readers who are just interested in the applications to representation theory can ignore all these examples except the ones in Section 2.3

Our main idea of applying the orbit hierarchy to the study of multiplicity diagrams was inspired by a short (unpublished) note of Duistermaat and Weinstein in which they point out that if $\mathcal{O}$ is a coadjoint orbit of $SU(n+1)$ that contains a point α in the interior of the positive Weyl chamber and α lies very close to some point on the ray

$$\lambda(n-1, -1, \ldots, -1), \quad \lambda > 0,$$

then $\mathcal{O}$ fibers equivariantly over $\mathbb{C}P^n$. This fact is used by them to compute the reduction of $\mathcal{O}$ with respect to the Cartan subgroup T of $SU(n+1)$ at any point β close to the origin in $\mathfrak{t}^*$, and, as an easy consequence of their computation, one can show that the Duistermaat–Heckman measure is equal to a constant multiple of Lebesque measure in a neighborhood of the origin. (This was *not* pointed out in their note, but was pointed out a short time afterward by Cushman and Duistermaat for the special case of the group $SU(3)$. Theirs was, in fact, the first observed example of lacunae. The tie-in between reduction and the Duistermaat–Heckman theorem will be described more carefully in the next chapter.)

2.1 Basic Example

Our first example is perhaps the basic example, in the sense that most of our other examples will be variants of it.

Example 2.1 Let

$$\rho: X \to B \tag{2.1}$$

be a differentiable fibration. Let $F_b = T^*(X_b)$ be the cotangent bundle of X_b, the fiber of X over $b \in B$. Then $T^*(X_b)$ is a symplectic manifold and so (2.1) gives a symplectic fibration

$$\pi: M \to B \tag{2.2}$$

whose fibers are the $T^*(X_b)$. Suppose we are given a connection A:

$$TX = Vert(X) \oplus Hor(X) \tag{A}$$

on X. Given a curve on B joining p to q, the connection A gives a diffeomorphism from X_p to X_q (at least locally), and hence a diffeomorphism of

$F_p = T^*X_p$ to $F_q = T^*X_q$. Thus the connection A on (2.1) gives rise to a connection Γ on (2.2). The action of Diff X_p on T^*X_p is Hamiltonian: Each vector field ξ on X_p corresponds to the function $-\langle\alpha_p, \hat\xi\rangle$ on T^*X_p, where α_p is the canonical one-form on T^*X_p and $\hat\xi$ is the lift of ξ to T^*X_p. Hence Γ has an associated coupling form ω_Γ (cf. the remark after Theorem 1.4.1). Let us describe the form ω_Γ associated to this connection: The splitting A gives rise to the dual splitting A^*:

$$T^*X = Vert^*(X) \oplus Hor^*(X). \qquad (A^*)$$

By definition $Hor^*(X)$ consists of those covectors that vanish on tangent vectors to the fibers of (2.1). Thus A^* gives an identification of $Vert^*X = Hor(X)^\circ$ with M, and hence an embedding

$$i_M: M \to T^*X.$$

Let α be the canonical one-form on T^*X and $\omega = d\alpha$ the canonical symplectic form on T^*X. It is clear that if we define

$$\alpha_\Gamma = i_M^*\alpha$$

then α_Γ, when restricted to each fiber F_b of (2.2), is just the canonical one-form α_b of $F_b = T^*X_b$. Hence $i_M^*\omega$ restricts to the canonical two-form of F_b. Let us now show that the orthogonal complement of $T(F_b)$ with respect to $i_M^*\omega$ is the fiber at b of the horizontal bundle of the connection on M. For this we recall our definition of the connection. Let v be a vector field on **B**. The connection on X then defines a horizontal vector field $v^\#$ on X. Let ϕ_t be the local one-parameter group generated by $v^\#$. Then ϕ_t carries fibers into fibers, and hence defines a one-parameter group of transformations ψ_t of M: If $\phi_t(X_a) = X_b$ then $F_a = T^*(X_a)$ and $F_b = T^*(X_b)$ and $\psi_t: F_a \to F_b$ is given by $\psi_t = (\phi_t^*)^{-1}$. If i_a denotes the embedding of F_a into M as the fiber over a, then it follows that

$$\psi_t^*(i_b^*\alpha_\Gamma) = i_a^*\alpha_\Gamma. \qquad (2.3)$$

The infinitesimal generator of ψ_t is the horizontal vector field $\tilde v$ associated to v. What we must show is that

$$[\iota(\tilde v)i_M^*\omega](w) = 0 \qquad (2.4)$$

for any vertical vector field w. Now

$$\iota(\tilde v)i_M^*\omega = \iota(\tilde v)d\alpha_\Gamma = L_{\tilde v}\alpha_\Gamma - d\iota(\tilde v)\alpha_\Gamma.$$

Equation (2.3) implies that

$$[L_{\tilde v}\alpha_\Gamma](w) = 0.$$

We claim that the function $\iota(\tilde{v})\alpha_\Gamma$ vanishes identically. Indeed, let us examine the value of this function at some point $z = (x, \xi)$ of M. Let π denote the projection of M and of T^*X onto X. Then $\pi \circ i_M = \pi$. By the definition of the canonical one-form, we have

$$\langle \alpha_\Gamma, \tilde{v} \rangle_z = \langle \xi', d\pi \circ di_M(\tilde{v})_z \rangle = \langle \xi', d\pi(\tilde{v})_z \rangle,$$

where $i_M(x, \xi) = (x, \xi')$. On the other hand, $d\pi(\tilde{v}) = v^\#$ is horizontal and $\xi' \in Vert^*X = Hor(X)^\circ$ so

$$\langle \xi', d\pi(\tilde{v})_z \rangle = 0.$$

We have thus proved the following basic theorem:

Theorem 2.1.1 *Let $X \to B$ be a fibration with connection. Let $M \to B$ be the fibration whose fiber over each $b \in B$ is $F_b = T^*(X_b)$. Then M is a symplectic fiber bundle with symplectic connection Γ. The connection on X induces a splitting $T^*X = Vert^*X \oplus Hor^*X$ and we have an identification of M with $Vert^*X$ and hence an embedding i_M of M into T^*X. If ω denotes the canonical two-form on T^*X then the form*

$$\omega_\Gamma = i_M^*\omega \tag{2.5}$$

is Γ-compatible.

Example 2.2 Here is a special case of Example 2.1: Let G be a Lie group and suppose that the X in Example 2.1 is a principal G-bundle and that A is a G-invariant connection defining a connection one-form $\theta_A : TX \to \mathfrak{g}$. Thus the presymplectic form ω_Γ of equation (2.5) is G-invariant. We claim that G acts on the presymplectic manifold M in a Hamiltonian fashion. (If the manifold is only presymplectic, a "Hamiltonian action" still means that one has a G-equivariant moment map $\Phi : M \longrightarrow \mathfrak{g}^*$ such that $\iota(\xi_M)\omega_\Gamma = -\langle d\Phi, \xi \rangle$.) We may identify $Vert(X)$ with a trivial bundle with fiber $\mathfrak{g}$, the Lie algebra of G, and hence

$$M \cong X \times \mathfrak{g}^* \cong Vert^* X. \tag{2.6}$$

In terms of the connection one-form θ_A the form $\alpha_\Gamma = i_M^*\alpha$ is given by $\alpha_\Gamma = \langle pr_2, \theta_A \rangle$, where $pr_2 : M \to \mathfrak{g}^*$ is the projection onto the second factor and $\langle \cdot, \cdot \rangle$ is the pairing between $\mathfrak{g}^*$ and $\mathfrak{g}$.

Let $\xi \in \mathfrak{g}$ and let ξ_M and ξ_X denote the corresponding vector fields on M and on X. Since α_Γ is G-invariant we have $0 = L_{\xi_M}\alpha_\Gamma = \iota(\xi_M)d\alpha_\Gamma + d\iota(\xi_M)\alpha_\Gamma$. But $d\alpha_\Gamma = \omega_\Gamma$ and $\iota(\xi_M)\alpha_\Gamma = \langle pr_2, \iota(\xi)\theta_A \rangle = \langle pr_2, \xi \rangle$. Thus $\iota(\xi_M)\omega_\Gamma = -d\langle pr_2, \xi \rangle$. This shows that pr_2 is a moment map for the G action.

The space $M = X \times \mathfrak{g}^*$ is sometimes referred to as the universal phase space of Weinstein. It is Weinstein's derivation of the minimal coupling construction of [S2] via reduction. The reason for the term "universal" is explained in the next example.

Example 2.3 As in example 2.2 let X be a principal G bundle over B, and let F be a manifold on which G acts. Let

$$X_F \to B \qquad (2.7)$$

be the corresponding associated bundle. Recall that $X_F = X \times_G F = (X \times F)/G$. Each fiber can be identified with F, but only up to the action of G. If F is a symplectic manifold and the action of G on F is symplectic, then X_F becomes a symplectic fiber bundle. Suppose that the action of G on F is Hamiltonian with moment map

$$\Phi_F \colon F \to \mathfrak{g}^*.$$

Choose a connection one-form on the principal bundle X. We can then form the bundle $M \cong X \times \mathfrak{g}^*$, as in Example 2.2, with connection and moment map. Consider the space $M \times F$. It is a Hamiltonian G-space with moment map $\Phi \colon M \times F \to \mathfrak{g}^*$ given by

$$\Phi(x, \beta, f) = \Phi_F(f) + \beta.$$

In particular $\Phi^{-1}(0)$ can be identified with $X \times F$ and hence

$$\Phi^{-1}(0)/G \cong X_F.$$

The form $\omega_\Gamma + \omega_F$ is G-invariant; hence so is its restriction to $\Phi^{-1}(0)$. Furthermore, for any $\xi \in \mathfrak{g}$ we have that $\iota(\xi_M)\omega_\Gamma + \iota(\xi_F)\omega_F = -d\langle\Phi_F + pr_2, \xi\rangle$ vanishes on $\Phi^{-1}(0)$. Hence the restriction of $\omega_\Gamma + \omega_F$ to $\Phi^{-1}(0)$ is the pullback of a presymplectic form $\omega_{\Gamma, F}$ on X_F. In other words, the space X_F inherits a presymplectic form $\omega_{\Gamma, F}$ whose restriction to the fiber is the symplectic form on the fiber, and whose associated connection is the connection associated to the given connection on the principal bundle X.

Explicitly, the presymplectic form $\omega_{\Gamma, F}$ at a point $[x, f] \in X \times_G F$ (where $[x, f]$ denotes the equivalence class of $(x, f) \in X \times F$, i.e., its G-orbit) can be written as follows. The tangent space $T_{[x, f]}(X \times_G F)$ is the direct sum of the vertical subspace and its symplectic perpendicular, the horizontal subspace. The vertical subspace is isomorphic to the tangent space $T_f F$ and the restriction of the form $\omega_{\Gamma, F}([x, f])$ to the vertical is isomorphic to $\omega_F(f)$. The horizontal subspace is isomorphic to the horizontal subspace of the tangent space $T_x X$ and the restriction of $\omega_{\Gamma, F}$ to this subspace is the two-form $-\langle\Phi(f), \Theta_A(x)\rangle$

obtained by pairing the value of the moment map at f with the curvature of the connection form θ_A at x, $\Theta_A(x)(X, Y) = d\theta_A(x)(X, Y)$ for any two horizontal vectors X and Y. Neither form depends on the choice of the pair (f, x) representing $[x, f]$ in the associated bundle.

Weinstein [W3] defined a connection θ_A to be *fat* at a point $\mu \in \mathfrak{g}^*$ if the two-form $\langle \mu, \Theta_A \rangle$ is nondegenerate on the horizontal subspace of the principal G-bundle X. It follows from this definition that the minimal coupling form ω_Γ on the space $M = X \times \mathfrak{g}^*$ is symplectic precisely on the open set $X \times U$, where

$$U = \{\mu \in \mathfrak{g}^* \mid \theta_A \text{ is fat at } \mu\}.$$

If F is a Hamiltonian G-space with the property that its image $\Phi_F(F)$ under the moment map is contained in the set U, then the associated bundle $X \times_G F$ is a symplectic manifold. In general it is next to impossible to determine explicitly the set of points at which a given connection is fat. However, it can be done in some special cases as we shall see in Section 2.3.

Suppose now that there is another group H acting on F in a Hamiltonian fashion with moment map $J: F \to \mathfrak{h}^*$ and that this action of H commutes with the action of G. Then the H moment map J is G-invariant. The action of H on F induces an action of H on the associated bundle $X_F = X \times_G F$. This induced action is Hamiltonian and the restriction of the induced moment map to the fiber is J.

Remark. The symplectic fibrations with connection constructed this way turn out to include all symplectic fiber bundles with connection for which the holonomy group is a finite-dimensional Lie group. In fact, let

$$Y \to B$$

be a symplectic fibration over a connected base B. Suppose that Γ is a symplectic connection on Y, let b_0 be a point of B, and let F be the fiber over b_0. We assume that the holonomy group over b_0 is the finite-dimensional Lie group G. If γ is a curve joining b_0 to some other point b of B, the connection gives a holonomy map[2]

$$\tau_\gamma: F_{b_0} \to F_b. \tag{2.8}$$

If γ_1 is another curve joining b_0 to b, then

$$\tau_{\gamma_1} = \tau_\gamma \circ g, \tag{2.9}$$

---

[2] Strictly speaking, for τ_γ to be defined we must make some assumptions about the integrability of horizontal vector fields on Y.

where $g \in G$. So we may let

$$X \to B$$

be the fiber bundle whose fiber over b is the set of all τ_γ of the form (2.8). By (2.9) X is a principal G-bundle. Furthermore every curve on B starting at b_0 has, by construction, a horizontal lift to X, so X is equipped with a G-invariant connection. One can then check that the original symplectic bundle with connection comes from the construction of Example 2.3.

Example 2.4 (base change) Let $\pi \colon X \to B$ be a symplectic fibration, let B_1 be a manifold, and let $f \colon B_1 \to B$ be a smooth map. Then we can pull X back to B_1 via f to obtain a fiber bundle $\pi_1 \colon X_1 \to B_1$ and a diagram

$$
\begin{array}{ccc}
X_1 & \xrightarrow{g} & X \\
\pi_1 \downarrow & & \downarrow \pi \\
B_1 & \xrightarrow{f} & B
\end{array}
\tag{2.10}
$$

If Γ is a connection on X we can pull it back to get a connection Γ_1 on X_1 which will be symplectic if Γ is. Furthermore the forms pull back consistently to give

$$\omega_{\Gamma_1} = g^* \omega_\Gamma. \tag{2.11}$$

Example 2.5 ([S2]) In Example 2.4 let us take $B_1 = T^*B$. Let ω_B be the standard symplectic form on T^*B. We can then add $\pi_1^* \omega_B$ to the form ω_{Γ_1} given by (2.9) to obtain the form

$$\omega_{\Gamma_1} + \pi_1^* \omega_B. \tag{2.12}$$

This is clearly closed and Γ_1-compatible. It is easy to show [S2] that (2.12) is always symplectic. The symplectic structure on X_1 defined by (2.12) is used in physics to adjoin "internal variables" to a classical dynamical system. (See [S2], [W2], [M1], [HPS], etc.)

2.2 A Normal Form Theorem

Example 2.6 (symplectic embedding) The construction of Example 2.3 can also be used to write the moment map in a simple form near fixed points of a Hamiltonian action of a Lie group G. If a fixed point is isolated and G is compact then the equivariant Darboux theorem implies that near the point the moment map is quadratic and that it is completely determined by the induced linear action on the tangent space at the point and the value of the symplectic form at the point (cf. [GS8], Theorem 32.1).

In general the set of fixed points of the action is a union of symplectic sub-manifolds. We claim that the moment map near such a fixed manifold B is quadratic in the normal directions and that it is determined by the induced linear action of the group on the fibers of the symplectic vector bundle $TB^{\perp}$, the *symplectic normal bundle* of B. (A typical fiber $(TB^{\perp})_p$ of this bundle is the symplectic perpendicular of $T_p B$ in the tangent space of the ambient manifold.)

We start by proving the following generalization of the Darboux theorem due to Alan Weinstein (cf. [W1]).

Theorem 2.2.1 (symplectic embedding) *Let (B, ω) be a symplectic manifold. Then there exists a one-to-one correspondence (modulo appropriate equivalences) between*

(1) symplectic vector bundles over B and

(2) symplectic embeddings of B into higher-dimensional symplectic manifolds.

Proof. One direction is easy. Given a symplectic embedding $i: B \rightarrow M$ we associate to it its symplectic normal bundle.

Conversely, suppose we are given a symplectic vector bundle $E \rightarrow B$, that is, E is a vector bundle and every fiber carries a linear symplectic structure qua vector space. The bundle E can also be thought of as an associated bundle for some principal bundle $G \rightarrow X \xrightarrow{\pi} B$, $E = X_V$. For example, we may put on E a Hermitian structure compatible with the symplectic form and take X to be the bundle of unitary frames of E. Then X is a principal $U(n)$-bundle. (To construct a Hermitian form on E, notice that the bundle $Sp(E)$ of Darboux frames of E is a principal $Sp(2n)$-bundle. The subgroup $U(n)$ acts on it from the right. Each $U(n)$-orbit corresponds to a Hermitian structure on a fiber. A smooth global section of the quotient bundle $Sp(E)/U(n)$ determines a Hermitian structure on E. Such a section exists because $Sp(2n)/U(n)$ is contractible.)

The structure group G acts symplectically on a typical fiber V of E. Recall that such action is Hamiltonian with moment map Φ_V being given by the formula

$$-\langle \xi, \Phi_V(v) \rangle = \frac{1}{2}\omega_V(\xi v, v),$$

where ω_V is the symplectic form on V and the ξ in the right-hand side of the equation is considered to be an element of $\mathfrak{gl}(V)$.

A choice of a connection one-form on X gives rise to a presymplectic form ω on $X_V = E$ whose restriction to the fiber is the symplectic form on the fiber (cf. Example 2.3). However, ω is not symplectic at the points of the zero section; in fact it vanishes there. Indeed, the inverse image of the zero section in $\Phi^{-1}(0)$ consists of all (x, β, f) with $\beta = 0$ and $f = 0$. The restriction of the form α_Γ to this submanifold vanishes (since $\beta = 0$); hence so does the restriction of ω_Γ and also the restriction of ω_F. Hence the restriction of ω to the zero section vanishes. We correct this by adding the pull-up of the symplectic form on the base B; the form $\Omega = \omega + \pi^*\omega_B$ is symplectic in a neighborhood of the zero section and its restriction to the zero section is simply ω_B. It is easy to see that the normal bundle to the embedding of B into (X_V, Ω) as the zero section is E. This proves the existence of a symplectic embedding with a given normal bundle.

It remains to prove the uniqueness, that is, to show that given two symplectic embeddings $i_1 : B \to M_1$ and $i_2 : B \to M_2$ there exist neighborhoods U_1 and U_2 of B in M_1 and M_2, respectively, and a symplectic diffeomorphism $\phi : U_1 \to U_2$ such that $\phi \circ i_1 = i_2$. We will omit the proof of this, referring the reader to [W2] for a similar argument.   □

Remark.   In complete analogy with the Darboux theorem, there exists an equivariant version of the symplectic embedding theorem: In the presence of a compact group of automorphisms H one has a one-to-one correspondence between symplectic H-vector bundles and symplectic embeddings of B fixed by H.

Let (M, ω) be a Hamiltonian H-space and (B, ω_B) a component of the fixed point set of H. Then H acts linearly on the fibers of the symplectic normal bundle ν of B. The action of the group, therefore, on a typical fiber V is Hamiltonian. Denote the resulting moment map by $\Psi_V \colon V \to \mathfrak{h}^*$. As in the proof of the symplectic embedding theorem we write the bundle ν as an associated bundle X_V for some principal bundle $G \to X \overset{\pi}{\to} B$ over the base B. Let Ψ denote the moment map for the H action on X_V, so the restriction of Ψ to a typical fiber is Ψ_V.

The equivariant symplectic embedding theorem asserts that there exist H-invariant neighborhoods U_0 of the zero section in X_V and U of B in M and an H-equivariant symplectic diffeomorphism $\phi \colon U \to U_0$ mapping B into the zero section. We conclude that $\Psi \circ \phi$ is an H moment map in a neighborhood of B in M. It is "quadratic" in the normal directions.

2.3   Fibrations of Coadjoint Orbits and the Cross Section Theorem

2.3.1   Introduction and the Basic Assumptions

In this section we consider a special case of Examples 2.2 and 2.3: The total space X of the principal bundle is a Lie group G, the base is a coadjoint orbit

$G \cdot \lambda$ of G (here "$\cdot$" denotes the coadjoint action), and the structure group is the isotropy group G_λ of λ. Our reason for studying this case is: If μ is a point in $\mathfrak{g}^*$ with the isotropy group contained in G_λ then, as manifolds, we have

$$G \cdot \mu = (G \times G_\lambda \cdot \mu)/G_\lambda = G \times_{G_\lambda} (G_\lambda \cdot \mu),$$

that is, the coadjoint orbit of μ is a fiber bundle associated with the principal G_λ-bundle $G_\lambda \to G \to G \cdot \lambda$. The main result of this section is that under a mild assumption that always holds when the group is compact the fibration of orbits $G \cdot \mu \to G \cdot \lambda$ is symplectic. Along the way we will also study the singularities of the minimal coupling form on $G \times \mathfrak{g}_\lambda^*$. We will relate them to the focal points of the embedding $G \cdot \lambda \hookrightarrow \mathfrak{g}^*$ of the coadjoint orbit in the dual of the Lie algebra. As a bonus we will get a new proof of the symplectic cross section theorem. Our main interest lies with compact groups. However, since we do not need the compactness as such, we will consider a more general case.

We now fix a Lie group G and a coadjoint orbit $G \cdot \lambda$ of G. We assume, first of all, that the coadjoint orbit is locally closed. That is to say, every point in the orbit has a neighborhood in $\mathfrak{g}^*$ such that the intersection of the orbit with the neighborhood is closed in the neighborhood. For example, a nilpotent orbit in $SL(2, \mathbb{R})$ is not closed, but it is locally closed. There is an example due to Mautner of a solvable Lie group with coadjoint orbits that are not locally closed [P].

Definition 2.3.1 *We say that a point λ in the dual of the Lie algebra $\mathfrak{g}^*$ is split if the Lie algebra $\mathfrak{g}_\lambda$ of the isotropy group G_λ of λ has a G_λ-invariant complement $\mathfrak{n}$ in $\mathfrak{g}$:*

$$\mathfrak{g} = \mathfrak{g}_\lambda \oplus \mathfrak{n} \quad (G_\lambda\text{-equivariant}). \tag{2.13}$$

Our second assumption is that the point λ that we fixed is split. What does this amount to?

1. Since the tangent space $T_\lambda(G \cdot \lambda)$ of the coadjoint orbit is canonically isomorphic to the annihilator $\mathfrak{g}_\lambda^\circ$ of the Lie algebra of G_λ in $\mathfrak{g}^*$, our assumption is equivalent to: *There exists a G_λ-invariant subspace $\mathfrak{n}$ in $\mathfrak{g}$ such that*

$$\mathfrak{g}^* = \mathfrak{g}_\lambda^\circ \oplus \mathfrak{n}^\circ = T_\lambda(G \cdot \lambda) \oplus \mathfrak{n}^\circ, \tag{2.14}$$

where $\mathfrak{g}_\lambda^\circ$ and $\mathfrak{n}^\circ$ are the annihilators in $\mathfrak{g}^*$ of the subspaces $\mathfrak{g}_\lambda$ and $\mathfrak{n}$, respectively.

2. Since the abstract normal bundle of the embedding of the orbit $G \cdot \lambda$ into $\mathfrak{g}^*$ is homogeneous, it is the vector bundle $G \times_{G_\lambda} (\mathfrak{g}^*/\mathfrak{g}_\lambda^\circ)$ associated with the principal G_λ-bundle $G_\lambda \to G \to G \cdot \lambda$. A G_λ-equivariant splitting (2.14)

allows us to identify the abstract normal bundle with the associated bundle $G \times_{G_\lambda} \mathfrak{n}^\circ$. This vector bundle has a natural G-equivariant "exponential map"

$$\mathcal{E}: G \times_{G_\lambda} \mathfrak{n}^\circ \to \mathfrak{g}^*, \quad [g, \eta] \mapsto g \cdot (\lambda + \eta). \qquad (2.15)$$

Here $[g, \eta]$ denotes the class of $(g, \eta) \in G \times \mathfrak{n}^\circ$ in the associated bundle $G \times_{G_\lambda} \mathfrak{n}^\circ$. Thus our assumption amounts to the existence of a G-invariant tubular neighborhood of the orbit $G \cdot \lambda$ in the dual of the Lie algebra. Note that it makes no sense to talk about a tubular neighborhood of an orbit that is not locally closed, hence the first assumption. When the group G is compact, the map $\mathcal{E}$ is the exponential map for a flat G-invariant metric on $\mathfrak{g}^*$.

3. When the group G is not compact, the preceding assertion about the existence of a tubular invariant neighborhood contains a small white lie. Although we know from the inverse function theorem that the map $\mathcal{E}$ is one-to-one on some neighborhood of the zero section of the bundle $G \times_{G_\lambda} \mathfrak{n}^\circ$, we don't know whether this neighborhood can be chosen to be G-invariant. On the other hand there exists a G-invariant neighborhood of the zero section consisting of the points at which the Jacobian of $\mathcal{E}$ is bijective. Indeed, since the map $\mathcal{E}$ is G-equivariant and since it is an embedding on each fiber, a point $[g, v]$ is a regular point of the map if and only if the orbit through the image $\mathcal{E}([g, v])$ is transverse to the image of the fiber $g \cdot (\lambda + \mathfrak{n}^\circ)$. That is to say, the set of regular points of $\mathcal{E}$ is $G \times_{G_\lambda} V_\lambda$, where

$$V_\lambda = \{[e, v] \mid G \cdot (\lambda + v) \text{ is transverse to } \lambda + \mathfrak{n}^\circ\}$$

and e denotes the identity element of G. And the set of regular values of $\mathcal{E}$ is $G \cdot \{x \in \lambda + \mathfrak{n}^\circ \mid G \cdot x \text{ is transverse to } \lambda + \mathfrak{n}^\circ\}$.

The map $\mathcal{E}$ restricted to the neighborhood $G \times_{G_\lambda} V_\lambda$ is a local diffeomorphism, a situation reminiscent of Luna's slice theorem. Also, for any vector w in the neighborhood, the map $\mathcal{E}: G \cdot w \to G \cdot \mathcal{E}(w)$ is a covering map (possibly one-to-one). Since the coadjoint orbits of compact groups are simply connected (see Subsection 2.3.2), the map $\mathcal{E}: G \cdot w \to G \cdot \mathcal{E}(w)$ is one-to-one for any compact Lie group G and any regular value w of $\mathcal{E}$.

4. The existence of a G_λ-equivariant splitting (2.13) is also equivalent to the existence of a G_λ-equivariant projection

$$A_0: T_e G = \mathfrak{g} \to \mathfrak{g}_\lambda, \qquad (2.16)$$

where e denotes the identity element of the group G. The space $\mathfrak{n}$ is then the kernel of the projection A_0. Translating A_0 around the group G by left multiplication, we get a $\mathfrak{g}_\lambda$-valued one-form $A \in \Omega^1(G, \mathfrak{g}_\lambda)$. It is easy to see that A is a connection form on the principal G_λ-bundle $G \to G \cdot \lambda$. Note that the canonical projection $\mathfrak{g}^* \to \mathfrak{g}_\lambda^*$, the transpose of the embedding $\mathfrak{g}_\lambda \hookrightarrow \mathfrak{g}$, identifies $\mathfrak{n}^\circ$ with $\mathfrak{g}_\lambda^*$. Conversely $A_0^\dagger$, the transpose of the projection A_0, maps $\mathfrak{g}_\lambda^*$ injectively onto $\mathfrak{n}^\circ$.

5. If the group G is compact, the case that is of most interest to us, then the existence of the splitting (2.13) is automatic at any point $\lambda \in \mathfrak{g}^*$. In general there may exist no such splitting at a given point λ even when the Lie algebra $\mathfrak{g}$ has an invariant inner product. One can show that for a reductive group a point $\lambda \in \mathfrak{g} \simeq \mathfrak{g}^*$ is split if and only if λ is semisimple.

To see how the splitting condition may fail take $G = SL(2, \mathbb{R})$ and consider the (co)adjoint orbit through $\lambda = \begin{pmatrix} 0 & 1 \\ 0 & 0 \end{pmatrix}$. Then $G_\lambda = \left\{ \begin{pmatrix} 1 & * \\ 0 & 1 \end{pmatrix} \right\}$ and it is easy to see that there exists no G_λ-equivariant splitting

$$\mathfrak{sl}(2, \mathbb{R}) = \mathbb{R} \begin{pmatrix} 0 & 1 \\ 0 & 0 \end{pmatrix} \oplus \mathfrak{n}.$$

2.3.2 Some Facts about Actions of Compact Groups

Recall that an action $G \times M \to M$, $(g, m) \mapsto g \cdot m$ of a Lie group G on a topological space M is called *proper* if the map $G \times M \to M \times M$, $(g, m) \mapsto (g \cdot m, m)$ is proper. The action of a compact Lie group is always proper. It is known that a proper action has slices at every point. A *slice* S at a point m for an action of a Lie group G on a manifold M is a G_m-invariant submanifold of M (G_m is the isotropy subgroup of m) such that the map

$$G \times_{G_m} S \to M, \quad [g, s] \mapsto g \cdot s$$

is a diffeomorphism onto its image and such that the image is an open neighborhood of the orbit $G \cdot m$. Two conditions together guarantee that a G_m-invariant submanifold S passing through m is actually a slice: (i) The tangent space to M at m is a direct sum of the tangent spaces to S and to the G-orbit, $T_m M = T_m S \oplus T_m G \cdot m$, and (ii) if s is a point in S and $a \cdot s$ is also in S for some $a \in G$ then $a \in G_m$.

Consider now the coadjoint action of a Lie group G. If G is not compact the action is not proper (the isotropy group of the origin is all of G hence not compact), and one shouldn't expect the existence of slices at every point. Note that the splitting condition (2.13) is necessary for the existence of a slice.

If the group G is compact, the coadjoint action is proper and slices do exist at every point. We will now discuss how to find these slices. The compactness of G guarantees the existence of an invariant inner product $(\cdot, \cdot)$ on $\mathfrak{g}$. The map $\mathfrak{g} \to \mathfrak{g}^*$, $X \mapsto (X, \cdot)$ is a G-equivariant isomorphism. Let us now fix an invariant inner product and a corresponding identification of $\mathfrak{g}$ and $\mathfrak{g}^*$. We next fix a Cartan subalgebra t of g. Recall that t is a maximal Abelian subalgebra and that it is unique up to the adjoint action of the elements of G (more facts about the structure of compact groups are reviewed in Chapter 5). It is well known that the space g/G of (co)adjoint orbits is isomorphic to a closed convex polyhedral cone t_+ in a Cartan subalgebra t, the cone being a (closed) Weyl chamber. Let λ be a point in $t_+ \subset g$ and let G_λ denote its isotropy group. It is not hard to see that t is also a Cartan subalgebra of $g_\lambda = \mathrm{Lie}\,(G_\lambda)$. However, the set $(t_\lambda)_+$, the subset of t that parameterizes the orbits of G_λ, is bigger than the Weyl chamber t_+ of G, $t_+ \subset (t_\lambda)_+$. In fact $(t_\lambda)_+$ can be chosen to be a union of t_+ and several of its translates by the elements of the Weyl group of G.

Lemma 2.3.2 *The open subset S_λ of g_λ defined by*

$$S_\lambda = G_\lambda \cdot \{x \in t_+ \mid G_x \subset G_\lambda\}$$

is a slice at λ for the coadjoint action of G. Moreover S_λ is the largest subset of g_λ that is a slice at λ.

Remarks

1. If G is connected (which we always tacitly assume) then the isotropy groups for its coadjoint action are connected as well. This fact could be deduced, for example, from simple-connectedness of coadjoint orbits and the long exact homotopy sequence. The simple-connectedness of (co)adjoint orbits, in turn, can be deduced via Morse theory. Namely, the distance function from an orbit to a generic point in the Lie algebra is a Morse function all of whose indices are even (cf. [Bo2]; more details can also be found in Section 2.3.4 to follow). Thus the inclusion $G_x \subset G_\lambda$ of isotropy groups is equivalent to the inclusion of the isotropy Lie algebras $g_x \subset g_\lambda$.

   It follows that the set $\{x \in t_+ \mid G_x \subset G_\lambda\} = \{x \in t_+ \mid g_x \subset g_\lambda\}$ is open in t_+, hence in $(t_\lambda)_+$. Consequently the set

   $$S_\lambda = G_\lambda \cdot \{x \in t_+ \mid g_x \subset g_\lambda\}$$

   is an open G_λ-invariant neighborhood of λ in g_λ.

2. The tangent space at λ of the orbit $G \cdot \lambda$ is orthogonal to g_λ relative to an invariant inner product on g. By dimension counting it is a full orthogonal complement of g_λ, $T_\lambda(G \cdot \lambda) = g_\lambda^\perp$. (Note that the splitting $g = g_\lambda \oplus g_\lambda^\perp$ is in fact independent of the choice of an inner product since $g_\lambda^\perp$ is the sum of root spaces of g corresponding to the roots that do not vanish on λ.)

It follows that $G \times_{G_\lambda} \mathfrak{g}_\lambda$ is a normal bundle for the embedding $G \cdot \lambda \hookrightarrow \mathfrak{g}$ and that $\mathcal{E}\colon G \times_{G_\lambda} \mathfrak{g}_\lambda \to \mathfrak{g}$, $[g, \nu] \mapsto g \cdot (\lambda + \nu)$ is the exponential map defined by the inner product.

Proof of Lemma 2.3.2. Suppose that x is a point in S_λ and that $g \cdot x$ is also in S_λ for some $g \in G$. We want to show that g is in the isotropy group G_λ. By definition of the set S_λ there are elements a and b in G_λ such that $a \cdot x$ and $b \cdot g \cdot x$ both lie in the Weyl chamber $\mathfrak{t}_+$ of G. But since the Weyl chamber parameterizes the orbits of G in one-to-one fashion we must have $a \cdot x = bg \cdot x$ whence $a^{-1} bg \cdot x = x$, which implies that $a^{-1} bg \in G_x \subset G_\lambda$, which implies that $g \in G_\lambda$. This proves that S_λ is a slice. $\square$

Example Consider $G = SO(3)$, the group of 3×3 special orthogonal matrices. The (co)adjoint representation of $SO(3)$ is isomorphic to its standard representation on $\mathbb{R}^3$. Let $\lambda = (1, 0, 0) \in \mathbb{R}^3$. Then its isotropy group G_λ is a circle, $\mathfrak{g}_\lambda = \mathbb{R}\lambda = \{(x_1, 0, 0) \mid x_1 \in \mathbb{R}\}$, and

$$G \times_{G_\lambda} \mathfrak{g}_\lambda \simeq S^2 \times \mathbb{R}.$$

With these identifications $\mathcal{E}\colon S^2 \times \mathbb{R} \to \mathbb{R}^3$ is given by $\mathcal{E}(x, t) = (1 + t)x$ (we think of the sphere S^2 as a subset of $\mathbb{R}^3$ and, if $x = g \cdot \lambda \in S^2$, $\eta = t\lambda \in \mathfrak{g}_\lambda$, identify $[g, \eta]$ with (x, tx)). Then $S_\lambda = \{(x_1, 0, 0) \mid x_1 > 0\}$. Note that whereas $\mathcal{E}$ maps $S^2 \times \{t > -1\}$ bijectively onto $\mathbb{R}^3 \setminus \{0\}$ the set of regular points of $\mathcal{E}$ is bigger: It is the set $S^2 \times \{t \neq -1\}$.

2.3.3 *Fibrations of Coadjoint Orbits*

The main result of this subsection, and of the section as a whole, is the following.

Theorem 2.3.3 *Let λ be a split point in $\mathfrak{g}^*$, let $\mathfrak{g} = \mathfrak{g}_\lambda \oplus \mathfrak{n}$ be a G_λ-equivariant splitting, and let $x \in \lambda + \mathfrak{n}^\circ$ be such that coadjoint orbit $G \cdot x$ is transverse to $\lambda + \mathfrak{n}^\circ$. Then, up to cover, the orbit $G \cdot x$ fibers symplectically over the coadjoint orbit $G \cdot \lambda$. The typical fiber is isomorphic to a coadjoint orbit of the isotropy group G_λ of λ: $G \times_{G_\lambda} G_\lambda \cdot x \simeq G \cdot x$. The symplectic connection on the fiber bundle equals the connection defined by the splitting of $\mathfrak{g}$ chosen above.*

Proof. Since the orbit $G \cdot x$ is transverse to the affine plane $\lambda + \mathfrak{n}^\circ$, x is a regular value of the "exponential map" $\mathcal{E}\colon G \times_{G_\lambda} \mathfrak{n}^\circ \to \mathfrak{g}^*$ defined in equation (2.15). Consequently the restriction of $\mathcal{E}$ to the orbit through a preimage $[e, x - \lambda] \in G \times_{G_\lambda} \mathfrak{n}^\circ$ of x, $\mathcal{E}\colon G \cdot [e, x - \lambda] \to G \cdot x$, is a (possibly one-to-one) covering map. Recall that if G is compact then $G \cdot x$ is simply connected, so $G \cdot [e, x - \lambda]$ is diffeomorphic to $G \cdot x$ in this case.

Since the coadjoint orbit $G \cdot x$ has a natural symplectic structure $\omega_{G \cdot x}$, its cover $G \cdot [e, x - \lambda]$ is also symplectic. On the other hand, the orbit $G \cdot [e, x - \lambda]$ is diffeomorphic to the associated bundle $G \times_{G_\lambda} G_\lambda \cdot (x - \lambda)$, whose base is the coadjoint orbit $G \cdot \lambda$.

To show that $G \cdot [e, x - \lambda] \to G \cdot \lambda$ is a symplectic fiber bundle it is enough to check that the tangent space $T_x(G_\lambda \cdot x)$ to the fiber is a symplectic subspace of the tangent space to the orbit $T_x(G \cdot x)$. Recall that the symplectic form $\omega_{G \cdot x}(x)$ on $T_x(G \cdot x)$ is defined by

$$\omega_{G \cdot x}(x)(\xi \cdot x, \eta \cdot x) = \langle x, \mathrm{ad}(\xi)\,\eta \rangle = -\langle \xi \cdot x, \eta \rangle$$

for any $\xi, \eta \in \mathfrak{g}$. Here and elsewhere $\langle , \rangle$ denotes the natural pairing between $\mathfrak{g}^*$ and $\mathfrak{g}$ and "$\cdot$" denotes the infinitesimal coadjoint action of $\mathfrak{g}$ on $\mathfrak{g}^*$, $\xi \cdot x = -\mathrm{ad}^\dagger(\xi)x$.

Recall that the transpose of the inclusion $i\colon \mathfrak{g}_\lambda \hookrightarrow \mathfrak{g}$ is a G_λ-equivariant projection $i^\dagger\colon \mathfrak{g}^* \to \mathfrak{g}_\lambda^*$ whose kernel is the annihilator $\mathfrak{g}_\lambda^\circ$. Since $\lambda + \mathfrak{n}^\circ$ is transverse to $\mathfrak{g}_\lambda^\circ$, $i^\dagger\colon \lambda + \mathfrak{n}^\circ \to \mathfrak{g}_\lambda^*$ is a G_λ-equivariant isomorphism which identifies the orbit $G_\lambda \cdot x$ with the coadjoint orbit $G_\lambda \cdot i^\dagger(x)$. Thus for any $X, Y \in \mathfrak{g}_\lambda$ we have

$$\begin{aligned}
\omega_{G \cdot x}(x)(X \cdot x, Y \cdot x) &= -\langle X \cdot x, Y \rangle \\
&= -\langle X \cdot i^\dagger(x), Y \rangle \quad \text{(since } i^\dagger \text{ is } G_\lambda\text{-equivariant)} \\
&= \omega_{G_\lambda \cdot i^\dagger(x)}(i^\dagger(x))(X \cdot i^\dagger(x), Y \cdot i^\dagger(x)).
\end{aligned}$$

This proves that $T_x(G_\lambda \cdot x)$ is a symplectic subspace of $(T_x(G \cdot x), \omega_{G \cdot x}(x))$.

It remains to show that the connection defined by the splitting $\mathfrak{g} = \mathfrak{g}_\lambda \oplus \mathfrak{n}$ and the symplectic connection are equal. For this it is enough to check that the horizontal subspaces defined by the splitting are symplectically perpendicular to the vertical subspaces. The map $\mathcal{E}$ maps the horizontal subspace at the point $[e, x - \lambda]$ to the subspace $\mathfrak{n} \cdot x := \{-\mathrm{ad}(\xi)^\dagger(x) \mid \xi \in \mathfrak{n}\}$. For any $\eta \in \mathfrak{g}_\lambda$ the vector $-\mathrm{ad}(\eta)^\dagger(x)$ lies in the annihilator of $\mathfrak{n}$ by the G_λ-invariance of the affine plane $\lambda + \mathfrak{n}^\circ$. Therefore, for any $\xi \in \mathfrak{n}$ and any $\eta \in \mathfrak{g}_\lambda$,

$$\omega_{G \cdot x}(x)(\xi \cdot x, \eta \cdot x) = \langle x, \mathrm{ad}(\xi)\,\eta \rangle = \langle \mathrm{ad}(\eta)^\dagger(x), \xi \rangle = 0. \qquad \square$$

Corollary 2.3.4 *Suppose G is compact and x and λ are two points in a Weyl chamber $\mathfrak{t}_+$ of $\mathfrak{g}$. If the isotropy Lie algebra $\mathfrak{g}_x$ of x is contained in $\mathfrak{g}_\lambda$ then the map of coadjoint orbits $G \cdot x \to G \cdot \lambda$, $g \cdot x \to g \cdot \lambda$, is a symplectic fibration with typical fiber isomorphic to a coadjoint orbit of G_λ. The symplectic* *connection on this fiber bundle equals the connection defined by the natural splitting* $\mathfrak{g} = \mathfrak{g}_\lambda \oplus \mathfrak{g}_\lambda^\perp$, *where* $\mathfrak{g}_\lambda^\perp$ *is the sum of all the root spaces corresponding to the roots that don't vanish on* λ.

To state the next corollary more succinctly we will pretend that the covering map $\mathcal{E}: G \times_{G_\lambda} G_\lambda \cdot x \to G \cdot x$ is one-to-one.

Corollary 2.3.5 *Let* λ *and* x *be as in Theorem 2.3.3. The symplectic form on the fiber bundle* $G \cdot x \to G \cdot \lambda$ *obtained by scaling by* ϵ *the fiber directions of the natural symplectic form on* $G \cdot x$ *equals the natural symplectic form on the coadjoint orbit through* $(1 - \epsilon)\lambda + \epsilon x$.

The following corollary is a version of the symplectic cross section theorem (cf. Theorem 26.7 in [GS8]).

Corollary 2.3.6 *Let* (M, ω) *be a Hamiltonian* G *space and* $\Phi: M \to \mathfrak{g}^*$ *a corresponding equivariant moment map. Suppose* λ *is a split point of* $\mathfrak{g}^*$, $\mathfrak{g} = \mathfrak{g}_\lambda \oplus \mathfrak{n}$ *is a splitting, and there is a* G_λ-*invariant neighborhood* W *of* λ *in the affine plane* $\lambda + \mathfrak{n}^\circ$ *with the property that* W *is a slice for the coadjoint action of* G. *Then the set* $R := \Phi^{-1}(W)$ *is a symplectic submanifold of* (M, ω) *and a Hamiltonian* G_λ-*space. The set* $G \cdot R$ *is an open subset of* M *which is* G-*equivariantly isomorphic to the associated bundle* $G \times_{G_\lambda} R$, *and the map* $G \times_{G_\lambda} R \to G \cdot \lambda$, $[g, m] \mapsto g \cdot \lambda$, *is a symplectic fibration (the symplectic form on* $G \times_{G_\lambda} R$ *comes from its identification with* $G \cdot R \hookrightarrow M$). *Finally, the symplectic connection on the bundle* $G \times_{G_\lambda} R \to G \cdot \lambda$ *equals the connection determined by the chosen splitting of* $\mathfrak{g}$.

Remarks

1. The submanifold R is called a *symplectic cross section*.
2. If G is compact, an invariant metric identifies the plane $\lambda + \mathfrak{n}^\circ$ with the isotropy Lie algebra $\mathfrak{g}_\lambda$. By Lemma 2.3.2 we may choose as a slice at λ the set

$$S_\lambda = G_\lambda \cdot \{x \in \mathfrak{t}_+ \mid \mathfrak{g}_x \in \mathfrak{g}_\lambda\},$$

where $\mathfrak{t}_+$ is a Weyl chamber of $\mathfrak{g}$ containing λ.
3. By G-equivariance of the moment map Φ, its restriction to the set $G \cdot R$ can be written as a map of fiber bundles

$$\Phi: G \times_{G_\lambda} R \to G \times_{G_\lambda} W_\lambda, \quad [g, m] \mapsto [g, \Phi(m)].$$

We will see in the course of the proof that $\Phi|_R$ can be interpreted as a G_λ moment map. This way of writing the moment map Φ is useful in studying singularities of moment maps and in writing down their normal forms.

Proof. Since W is a slice, for any $w \in W$ the coadjoint orbit through w is transverse to W. Since the moment map Φ is equivariant it is transverse to W. Therefore R is a submanifold.

Since W is a slice, the set $G \cdot W$ is an open subset of $\mathfrak{g}^*$ diffeomorphic to the associated bundle $G \times_{G_\lambda} W$. Since the moment map Φ is equivariant, the set $G \cdot R$ is an open subset M diffeomorphic to $G \times_{G_\lambda} R$.

Let m be a point in the cross section R and let $x = \Phi(m)$. Consider the horizontal space at m of the connection associated with the splitting of $\mathfrak{g}$. It is the space $\mathfrak{n}_M(m) := \{\xi_M(m) \mid \xi \in \mathfrak{n}\}$, where ξ_M denotes the vector field on M defined by the infinitesimal action of $\xi \in \mathfrak{g}$. Now for any ξ and η in $\mathfrak{n}$ we have

$$
\begin{aligned}
\omega(m)(\xi_M(m), \eta_M(m)) &= -\langle \xi, d\Phi_m(\eta_M(m)\rangle \\
&= -\langle \xi, \eta \cdot x\rangle \\
&= -\omega_{G \cdot x}(\xi \cdot x, \eta \cdot x).
\end{aligned} \tag{2.17}
$$

But as we saw in the proof of Theorem 2.3.3 the pairing

$$
\mathfrak{n} \times \mathfrak{n} \to \mathbb{R}, \quad (\xi, \eta) \mapsto \omega_{G \cdot x}(x)(\xi \cdot x, \eta \cdot x) \tag{2.18}
$$

is nondegenerate for any $x \in \lambda + \mathfrak{n}^\circ$ with the property that the orbit $G \cdot x$ is transverse to $\lambda + \mathfrak{n}^\circ$. Therefore that space $\mathfrak{n}_M(m)$ is a symplectic subspace of the tangent space $T_m M$. For any vector v in the tangent space to the fiber $T_m R$ and any $\xi \in \mathfrak{n}$ we have

$$
\omega(m)(\xi_M(m), v) = -\langle \xi, d\Phi_m(v)\rangle = 0,
$$

since $\xi \in \mathfrak{n}$ and $d\Phi_m(v)$ is tangent to the affine plane $\lambda + \mathfrak{n}^\circ$, hence lies in $\mathfrak{n}^\circ$. Since we also have that $T_m M = T_m R \oplus \mathfrak{n}_M(m)$, the space $T_m R$ is the symplectic perpendicular of the symplectic subspace $\mathfrak{n}_M(m)$, whence $T_m R$ is itself symplectic. This proves that R is a symplectic submanifold of (M, ω).

Since the transpose of the inclusion $i \colon \mathfrak{g}_\lambda \to \mathfrak{g}$ identifies the affine plane $\lambda + \mathfrak{n}^\circ$ G_λ-equivariantly with $\mathfrak{g}_\lambda^*$, since R is a G_λ-invariant submanifold of (M, ω), and since the action of G on (M, ω) is Hamiltonian, it follows that the restriction of the G moment map Φ to R followed by the projection $i^\dagger$ shifted by λ is a moment map Φ_R for the action of G_λ on R, $\Phi_R = i^\dagger \circ (\Phi|_R - \lambda)$.

It remains to show that the symplectic connection on the fiber bundle $G \times_{G_\lambda} R \to G \cdot \lambda$, $[g, m] \to g \cdot \lambda$, equals the connection determined by the splitting $\mathfrak{g} = \mathfrak{g}_\lambda \oplus \mathfrak{n}$. But we have already seen that for any $v \in T_m R$ and any $\xi \in \mathfrak{n}$ we have $\omega(m)(\xi_M(m), v) = 0$. $\square$

The symplectic form ω on the fiber bundle $G \times_{G_\lambda} R \to G \cdot \lambda$ could be reconstructed from its restriction to the fiber $\omega_R = \omega|_R$ and the connection corresponding to the splitting $\mathfrak{g} = \mathfrak{g}_\lambda \oplus \mathfrak{n}$. To reconstruct we simply follow the prescription described in Example 2.2. We form the product of R with the presymplectic manifold $M = G \times \mathfrak{g}_\lambda^*$ and then reduce it at zero with respect to the Hamiltonian action of G_λ. The presymplectic form ω_Γ on M is chosen just as in Example 2.2 with one modification: We add the pullback of the natural symplectic form on the base $\omega_{G\cdot\lambda}$,

$$\omega_\Gamma = d\langle pr_2, A \rangle + \pi^* \omega_{G\cdot\lambda}.$$

Here A is the $\mathfrak{g}_\lambda$-valued one-form on G obtained by left translating the projection $A_0: \mathfrak{g} \to \mathfrak{g}_\lambda$ (cf. equation (2.16)). A computation using equation (2.17) verifies that our choice of ω_Γ is correct. Recall that the minimal coupling form ω_Γ is uniquely defined up to a pullback of a form from the base (Theorem 1.4.3).

Notice also that the preceding proof shows the following. Consider the neighborhood of λ in the affine plane $\lambda + \mathfrak{n}^\circ$ given by

$$\{x \in \lambda + \mathfrak{n}^\circ \mid G \cdot x \text{ is transverse to } \lambda + \mathfrak{n}^\circ\}.$$

Then the preimage under the moment map of this neighborhood is a symplectic G_λ-invariant submanifold of (M, ω) and a Hamiltonian G_λ-space. These two observations suggest a theorem (cf. Examples 2.2 and 2.3).

Theorem 2.3.7 *Let λ be a split point in $\mathfrak{g}^*$. Let A be a G-invariant connection form on the principal G_λ-bundle $\pi: G \to G \cdot \lambda$ corresponding to a choice of a splitting $\mathfrak{g} = \mathfrak{g}_\lambda \oplus \mathfrak{n}$. Then the $(G \times G_\lambda)$-invariant closed two-form ω_Γ on $G \times \mathfrak{g}_\lambda^*$ given by*

$$\omega_\Gamma = \pi^* \omega_{G\cdot\lambda} + d\langle pr_2, A \rangle$$

is nondegenerate at a point $(g, v) \in G \times \mathfrak{g}_\lambda^*$ if and only if the coadjoint orbit $G \cdot (\lambda + A_0^\dagger v)$ is transverse to the affine plane $\lambda + \mathfrak{n}^\circ$. Here $A_0^\dagger: \mathfrak{g}_\lambda^* \to \mathfrak{g}^*$ is the transpose of the projection $A_0: \mathfrak{g} \to \mathfrak{g}_\lambda$ defined by the splitting chosen (cf. equation (2.16)).

Proof. The proof is a computation combined with the observation made in the course of the proof of Theorem 2.3.3 that for $x \in \lambda + \mathfrak{n}^\circ$ with the property that the orbit $G \cdot x$ is transverse to the plane $t\lambda + \mathfrak{n}^\circ$ the pairing

$$\mathfrak{n} \times \mathfrak{n} \to \mathbb{R}, \quad (X, Y) \to \langle x, [X, Y] \rangle$$

is nondegenerate.

Note that the converse of the observation holds as well. Indeed suppose that the intersection is not transverse, that is, $\mathfrak{g}^* \neq T_x(G \cdot x) + T_x(\lambda + \mathfrak{n}^\circ) = \mathfrak{g}_x^\circ + \mathfrak{n}^\circ$. Taking the annihilators of both sides we get that $0 \neq \mathfrak{g}_x \cap \mathfrak{n}$. So there is a nonzero vector $X \in \mathfrak{n}$ that also lies in the isotropy Lie algebra of x. Therefore for any $Y \in \mathfrak{n}$ we have

$$0 = \langle \mathrm{ad}(X)^\dagger x, Y \rangle = \langle x, [X, Y] \rangle,$$

and so the pairing is degenerate.

It remains to show that the form ω_Γ is nondegenerate at a point $(g, \nu) \in G \times \mathfrak{g}_\lambda^*$ if and only if the pairing

$$\mathfrak{n} \times \mathfrak{n} \to \mathbb{R}, \quad (X, Y) \to \langle \lambda + A_0^\dagger \nu, [X, Y] \rangle$$

is nondegenerate. Since ω_Γ is $(G \times G_\lambda)$-invariant and since $G \times \mathfrak{g}_\lambda^* \to G \cdot \lambda$ is a symplectic fibration, the form ω_Γ is nondegenerate at a point (g, ν) if and only if its restriction to the horizontal subspace at a point (e, ν) is nondegenerate. Here, again, e denotes the identity of the group G. The tangent space of $G \times \mathfrak{g}_\lambda^*$ at (e, ν) is $\mathfrak{g} \times \mathfrak{g}_\lambda^*$ and the horizontal subspace is $\mathfrak{n} \times \{0\}$. The curvature of the connection form A at e is given by $(X, Y) \mapsto A_0([X, Y])$ for any $X, Y \in \mathfrak{n}$. And the value of the pullback of the natural form on $G \cdot \lambda$ at the point e on vectors $X, Y \in \mathfrak{n}$ is given by $(\pi^* \omega_{G \cdot \lambda})(e)(X, Y) = \langle \lambda, [X, Y] \rangle$. We conclude that ω_Γ is nondegenerate at (g, ν) if and only if the pairing $\mathfrak{n} \times \mathfrak{n} \to \mathbb{R}$ given by

$$(X, Y) \mapsto \langle \lambda, [X, Y] \rangle + \langle \nu, A_0([X, Y]) \rangle = \langle \lambda + A_0^\dagger \nu, [X, Y] \rangle$$

is nondegenerate and we are done. $\qquad\square$

If the group G is compact Theorem 2.3.7 takes the following form.

Corollary 2.3.8 *Suppose G is a compact group and λ is a point in $\mathfrak{g}$. Then the G-invariant connection form A on the principal G_λ-bundle $G \to G \cdot \lambda$ defined by the canonical splitting $\mathfrak{g} = \mathfrak{g}_\lambda \oplus \mathfrak{g}_\lambda^\perp$ is fat at a point $\nu \in \mathfrak{g}_\lambda$ if and only if the adjoint orbit $G \cdot \nu$ fibers over the adjoint orbit through λ.*

Consequently, if one coadjoint orbit of a compact group fibers over another orbit then it fibers symplectically.

Here is an example of a fibration of one orbit by another that does not satisfy condition (2.13), and for which the conclusion of Theorem 2.3.3 does not hold. This example was kindly provided to us by Prof. Schmid. Let $G = SL(3, \mathbb{R})$, so $\mathfrak{g}$ consists of all 3×3 matrices with trace 0. We may identify $\mathfrak{g}$ with $\mathfrak{g}^*$ using the Killing form (which is just the bilinear from $b(A, B) = \operatorname{tr} AB$). Let $\mathcal{O}_1$ consist of all nilpotent matrices of rank 2. So $\mathcal{O}_1$ is the orbit through the matrix

$$\beta = \begin{pmatrix} 0 & 1 & 0 \\ 0 & 0 & 1 \\ 0 & 0 & 0 \end{pmatrix}.$$

It is six-dimensional. Let $\mathcal{O}$ consist of all rank-1 nilpotent matrices, so $\mathcal{O}$ is the orbit through the matrix

$$\alpha = \begin{pmatrix} 0 & 0 & 1 \\ 0 & 0 & 0 \\ 0 & 0 & 0 \end{pmatrix}.$$

It is four-dimensional. The map $A \rightsquigarrow A^2$ gives a G-equivariant fibration $\mathcal{O}_1 \to \mathcal{O}$. A direct computation (which we leave to the reader) shows that (2.13) does not hold at any point of $\mathcal{O}$ and that the fibers of $\mathcal{O}_1 \to \mathcal{O}$ are isotropic submanifolds.

2.3.4 Morse Theory

In this subsection we describe the critical values of the exponential map and, using Morse theory, apply these results to the topology of the coadjoint orbits.

Let $\mathcal{O} = K \cdot \alpha$ be a coadjoint orbit through the split point α. Thus the isotropy subalgebra $\mathfrak{h} = \mathfrak{k}_\alpha$ has an $\mathfrak{h}$-invariant complement $\mathfrak{m}$ so that

$$\mathfrak{k} = \mathfrak{h} \oplus \mathfrak{m}$$

and hence

$$\mathfrak{k}^* = \mathfrak{h}^* + \mathfrak{m}^*.$$

Here $\mathfrak{m}^* = \mathfrak{h}^0$ can be identified with the tangent space to the orbit at α and hence $\mathfrak{h}^* = \mathfrak{m}^0$ is identified with the normal space. This means that we have an identification of the normal bundle as an associated bundle to the principal H-bundle $K \to K/H$ given by

$$N(\mathcal{O}) = K \times_H \mathfrak{h}^*.$$

The exponential map

$$E \colon N(\mathcal{O}) \to \mathfrak{k}^*$$

is defined by

$$E(\beta, \mu) = \beta + \mu,$$

where $\mu \in N(\mathcal{O})_\beta$, $\beta \in \mathcal{O}$, is regarded as an element of $\mathfrak{k}^*$. We would like to identify the focal points which are, by definition, the critical values of E. The main result due to Bott [Bo1] is the following:

Proposition 2.3.9 *Let $\gamma = E(\beta, \mu)$, $\gamma \neq \beta$, be a regular point of $\mathfrak{k}^*$ in the sense that the coadjoint orbit through γ has maximal dimension, $\dim k - r$. Then a point $\gamma_t = E(\beta, t\mu) = \beta + t\mu$ on the (normal) line through β and γ is a focal point if and only if the coadjoint orbit through γ_t is not regular. Furthermore, the corank of the differential of the exponential map is then given by*

$$\operatorname{corank} d E_{(\beta, t\mu)} = \dim K \cdot \gamma - \dim K \cdot \gamma_t.$$

In particular, this corank is always even.

This proposition has immediate consequences for the computation of the homology of the orbit $\mathcal{O}$ in the case where we have a positive-definite invariant scalar product on $\mathfrak{k}^*$ and the splitting and exponential map are relative to this scalar product. For example, if we choose γ to be a regular point not on the orbit $\mathcal{O}$, then the distance function from γ to $\mathcal{O}$ is a Morse function whose critical points are the points $\beta \in \mathcal{O}$ at which the line from γ to β is perpendicular to $\mathcal{O}$. The index of the distance function at β is the sum of the coranks of the focal points between β and γ. By the proposition, these indices are all even. We conclude that $\mathcal{O}$ has homology only in even dimensions and no torsion. We will return to more detailed use of this method for the computation of cohomology later on. In the meanwhile, let us prove Proposition 2.3.9.

Let π denote the projection $\pi: K \times \mathfrak{h}^* \to K \times_H \mathfrak{h}^* = N(\mathcal{O})$, and let $F = E \circ \pi$ so that

$$F(a, v) = a \cdot (\alpha + v).$$

It is enough to prove the proposition for the case that $\beta = \alpha$ and hence we want to compute dF at a point (e, v) where $v = tv_0$ and $\gamma = F(e, v_0)$ is regular. Identifying the tangent space $T K_e$ with the Lie algebra $\mathfrak{k}$, we clearly have

$$d F_{e,v}(\xi, \mu) = \xi \cdot (\alpha + v) + \mu.$$

Thus if $\eta \in \mathfrak{h}$ then

$$\{(\eta, -\eta \cdot v)\} \in \ker d F_{e,v},$$

corresponding to the action of $\mathfrak{h}$, and the point $[(e, v)]$ is a critical point of E if and only if the kernel is strictly larger. So if

$$\xi \cdot (\alpha + v) = 0 \tag{2.19}$$

and

$$\xi \cdot \alpha \neq 0,$$

then $[(e, \nu)]$ is critical for E. Thus the condition (2.19), which is, a priori, weaker than

$$\xi \cdot \alpha = 0, \quad \xi \cdot \nu = 0, \tag{2.20}$$

is equivalent to (2.20) for noncritical points of E. Now all points of the form $\pi(e, t\nu_0) = (\alpha, \nu)$, $t \neq 0$, are not critical for E if $|t|$ is sufficiently small, since the exponential map is regular near the zero section. Also all points of the form $\alpha + \nu = E(\alpha, \nu)$ are generic points of $\mathfrak{k}^*$ for sufficiently small $|t|$ since the line through α and $\alpha + \nu_0$ contains regular points, and hence all but a finite number of points on this line are regular, as nonregularity is a polynomial condition. Hence for all such points, condition (2.19) is equivalent to condition (2.20), which, in turn, is equivalent to

$$\xi \cdot \alpha = 0, \quad \xi \cdot \nu_0 = 0. \tag{2.21}$$

In particular, the dimension of the space of solutions of (2.21) is r, and a point $\alpha + \nu$ on our line is regular if and only if (2.19) is equivalent to (2.21). On the other hand, for $|t|$ small, $t \neq 0$, we have

$$\{(\xi, \mu) \mid \mu = -\xi \cdot (\alpha + t\nu_0)\} = \{(\xi, \mu) \mid \xi \cdot \alpha = 0, \ \mu = -t\xi \cdot \nu_0\},$$

since all these points are noncritical. In other words, for all noncritical points

$$\mathfrak{n}^0\{\xi \cdot (\alpha + t\nu_0)\} = \{\eta \cdot \nu_0 \mid \eta \in \mathfrak{h}\}.$$

But the right-hand side is independent of t and the subspace described by the left-hand side depends continuously on t. Hence this equality holds for all t. Thus, for $t \neq 0$, a solution of

$$\mu = -\xi \cdot (\alpha + t\nu_0), \quad \xi \cdot \alpha \neq 0,$$

with $\mu \in \mathfrak{n}^0$ exists if and only if a solution of

$$\xi \cdot (\alpha + t\nu_0) = 0, \quad \xi \cdot \alpha \neq 0,$$

exists, if and only if the point $\alpha + t\nu_0$ is not regular.

Suppose that we specialize to the case where $\mathfrak{k}$ has an invariant (nondegenerate) scalar product $(\,,\,)$ which we use to identify $\mathfrak{k}$ with $\mathfrak{k}^*$. So coadjoint orbits become adjoint orbits. Let us use the scalar product to choose the invariant complement. So

$$T(K \cdot \alpha)_\alpha = \mathfrak{h}^0 = \mathfrak{h}^\perp = \{[\xi, \alpha]\},$$

and
$$N(K \cdot \alpha)_\alpha = \mathfrak{h} = \{\eta \mid [\eta, \alpha] = 0\}.$$

Notice that these spaces are orthogonal complements since the invariance of $(\,,\,)$ implies that
$$([\xi, \alpha], \eta) = (\xi, [\alpha, \eta])$$
and $(\,,\,)$ is nondegenerate.

Proposition 2.3.10 *If a line is perpendicular to an orbit through one of its points, then it is perpendicular to the orbit through any of its points.*

Proof. The line $\{\alpha + t\nu\}$ is perpendicular to the orbit through α if and only if $[\nu, \alpha] = 0$ by the preceding remark. But then $[\nu, \alpha + t\nu] = 0$ for all t since $[\nu, \nu] = 0$. □

Suppose, now, that $\mathfrak{k}$ is compact and that $\mathfrak{t}$ is a Cartan subalgebra of $\mathfrak{k}$, so $\mathfrak{t}$ is a maximal abelian subalgebra, and let $\mathfrak{t}_{\mathrm{reg}} \subset \mathfrak{t}$ consist of the regular elements. Thus every orbit intersects $\mathfrak{t}$, and the regular orbits intersect $\mathfrak{t}$ in $\mathfrak{t}_{\mathrm{reg}}$. Furthermore, the complement of $\mathfrak{t}_{\mathrm{reg}}$ in $\mathfrak{t}$ is the union of a finite number of hyperplanes H_j. If $\nu \in \mathfrak{t}_{\mathrm{reg}}$ then the centralizer of ν is $\mathfrak{t}$. In other words, the normal to the orbit through ν is $\mathfrak{t}$. If $\mathcal{O}$ is any other orbit and the line from ν to $\alpha \in \mathcal{O}$ is perpendicular to $\mathcal{O}$ at α, then this line must be perpendicular to the orbit through ν by the previous proposition, and hence must lie in $\mathfrak{t}$. We have thus proved:

Proposition 2.3.11 *Let ν be a point of $\mathfrak{t}_{\mathrm{reg}}$, let $\mathcal{O}$ be a (co)adjoint orbit, and let $\ell_{\nu, \mathcal{O}}$ be the function on $\mathcal{O}$ given as the distance from ν. Then the critical points of $\ell_{\nu, \mathcal{O}}$ are the intersection points of $\mathcal{O}$ with $\mathfrak{t}$. In particular, these critical points are independent of the choice of ν in $\mathfrak{t}_{\mathrm{reg}}$.*

The (co)adjoint orbits in this case are called generalized flag manifolds. The preceding three propositions and a bit of Morse theory then lead to a purely algebraic computation of the homology of generalized flag manifolds that the senior authors learned from Bott many years ago: Let $\mathcal{O}$ be an orbit. Pick a point $\nu \in \mathfrak{t}_{\mathrm{reg}}$. The critical points of $\ell_{\nu, \mathcal{O}}$ are the points of intersection of $\mathcal{O}$ with $\mathfrak{t}_{\mathrm{reg}}$. These points can be determined by solving algebraic equations. (For example, if $\mathfrak{g} = \mathfrak{u}(n)$, then the problem reduces to finding the (common) eigenvalues of an element of $\mathcal{O}$.) Let these points be $\{\alpha_1, \ldots \alpha_k\}$. Each of these critical points has even index given by the sum over β of $r - \dim K \cdot \beta$, where β ranges over the points of intersection of the segment ν, α_k. But this sum is easily seen to be twice the number of hyperplanes H_i intersected by this segment. Since there is no homology in odd dimensions, the pth Betti number is the number of critical points of index p.

Here are some simple illustrations of the preceding propositions: If we take $K = SO(3)$ (or $SU(2)$) then $\mathfrak{k} = \mathbb{R}^3$ and $\mathfrak{t}$ is a line. The regular points of $\mathfrak{t}$ are the nonzero points. The coadjoint orbits are spheres, except for the single zero-dimensional coadjoint orbit consisting of the origin. For any sphere, the only focal point is the origin. If $\nu \neq 0$ is a point not on the sphere $\mathcal{O}$, then there is one perpendicular line from ν to $\mathcal{O}$ that intersects $\mathcal{O}$ in two points. The segment joining the closer point to ν does not contain the origin. The other point of intersection corresponds to the furthest point on $\mathcal{O}$ from ν. The homology of the sphere is $\mathbb{Z}$ in dimensions 0 and 2.

Suppose we take $K = U(3)$, so $\mathfrak{k} = \mathfrak{u}(3)$ is nine-dimensional, and we may take $\mathfrak{t}$ to consist of diagonal elements (with imaginary entries along the diagonal), so $\mathfrak{t}$ is three-dimensional. A regular element is one with distinct eigenvalues. The regular orbits are six-dimensional (six = nine $-$ three). If $\mathcal{O}$ is a regular orbit and $\alpha \in \mathcal{O}$ then α determines three orthogonal lines in $\mathbb{C}^3$, the eigendirections of α. Conversely, given an ordered set of three orthogonal lines, there is exactly one $\alpha \in \mathcal{O}$. So each regular orbit is diffeomorphic to the space of three ordered orthogonal lines in $\mathbb{C}^3$. The symplectic structure depends on the choice of $\mathcal{O}$. The orbit through α is four-dimensional if exactly two of the eigenvalues of α are equal and the third is distinct. Given a four-dimensional orbit $\mathcal{O}$, an $\alpha \in \mathcal{O}$ determines and is determined by a line, the eigendirection of the third eigenvalue. (Of course this line determines its orthogonal plane.) Thus the four-dimensional orbits are diffeomorphic to $\mathbb{C}P^2$, complex projective two space. The symplectic structure depends on the choice of $\mathcal{O}$. There are also zero-dimensional orbits (points) consisting of multiples of the identity matrix.

More generally, this computation works for $U(n)$ and shows that each orbit corresponds to an ordered collection of orthogonal subspaces of dimensions $n_1 \leq n_2 \leq \cdots \leq n_k$ with $n_1 + n_2 + \cdots + n_k = n$; in other words each orbit corresponds to the space of flags of a given type (hence the terminology of "generalized flag manifolds").

Returning to $U(3)$, let us write $\Delta(a, b, c)$, where $a, b, c \in \mathbb{R}$, for the diagonal matrix whose entries are (ia, ib, ic). There are three singular hyperplanes on which (at least) two of the eigenvalues are equal, $H_{a=b}$, $H_{a=c}$, $H_{b=c}$. Let us examine the four-dimensional orbits, say the orbit through $\alpha = \Delta(1, 1, 0)$. This orbit intersects $\mathfrak{t}$ in two other points: $\beta = \Delta(1, 0, 1)$ and $\gamma = \Delta(0, 1, 1)$. According to the recipe, we must pick a point in $\mathfrak{t}_{\text{reg}}$; suppose we pick

$\nu = \Delta(1, 2, 3)$. The segment from ν to γ consists of all points of the form

$$\Delta(t, 1 + t, 1 + 2t), \quad 0 < t < 1.$$

All such points are regular. Hence γ is a point of index zero. The segment from ν to β consists of all points of the form

$$\Delta(1, 2t, 1 + 2t).$$

There is one singular point at $t = \frac{1}{2}$. So β is a point of index two. The segment from ν to α consists of all points of the form

$$\Delta(1, 1 + t, 3t).$$

This segment crosses the hyperplane $H_{a=c}$ at $t = \frac{1}{3}$ and crosses the hyperplane $H_{b=c}$ at $t = \frac{1}{2}$. Hence α is a point of index four. Hence the homology of $\mathbb{C}P^2$ is $\mathbb{Z}$ in dimensions 0, 2, and 4 and zero in odd dimensions. A similar computation gives the homology of $\mathbb{C}P^n$ for any n.

2.4 Symplectic Mackey–Wigner Theory

Example 2.7 Let V be a vector space and B a submanifold of V. The abelian group V^* acts on T^*B as follows. A $w \in V^*$ is, by definition, a linear function on V. Let f_w denote the restriction of this function to B and let $\tau_w : T^*B \to T^*B$ be the symplectic map sending (b, β) into $(b, \beta + df_w(b))$. This defines a Hamiltonian action of V^* on T^*B.

Now let K be a compact Lie group and suppose that we are given a representation ρ of K on V and that B is an orbit of K in V. So K acts on B and this action lifts to a Hamiltonian action of K on T^*B. The actions of V^* and K fit together to give an action of the semidirect product

$$H = K \ltimes V^*$$

on T^*B. This action is transitive. So (up to covering) T^*B has to be a coadjoint orbit of H. Identifying the Lie algebra of H with the semidirect sum $\mathfrak{k} \oplus V^*$ and its dual with $\mathfrak{k}^* \oplus V$, one can easily see that this orbit is the orbit containing $\{0\} \times B$ and that T^*B is, in fact, H-equivariantly equivalent to this orbit.

Now fix a point b_0 of B and let $G = K_{b_0}$ be its isotropy subgroup. We can then identify $B = K/G$, and so K becomes a principal G-bundle over B as in subsection 2.3.1. Let F be a Hamiltonian G-space and construct the corresponding symplectic fibration $K_F \to B$. The discussion in Subsection 2.3.1 showed that a splitting (2.13) gives rise to a connection Γ on this bundle and hence, by minimal coupling, to a presymplectic form $\omega_{\Gamma,F}$. Moreover the natural action of K on K_F preserves $\omega_{\Gamma,F}$ and is a Hamiltonian action. Extend this action to a Hamiltonian action of H by letting the normal subgroup V^* act trivially. Pull the fibration $K_F \to B$ back to T^*B, and let $\mathcal{O}_F$ denote the pulled-back bundle. So we get the diagram

$$
\begin{array}{ccc}
\mathcal{O}_F & \longrightarrow & K_F \\
\downarrow & & \downarrow \\
T^*B & \longrightarrow & B
\end{array}
$$

where the top and the left arrows are H-morphisms. Thus $\mathcal{O}_F$ is an H-space and if we equip it with the symplectic form (2.11) it becomes a Hamiltonian H-space (since both T^*B and K_F are Hamiltonian H-spaces). Moreover H acts transitively on $\mathcal{O}_F$, so (up to a covering) $\mathcal{O}_F$ must be a coadjoint orbit of H. In fact, if we let i be the injection dual to the projection $\mathfrak{k} \to \mathfrak{g}$ then $(i(F), b_0)$ is an orbit of G in $\mathfrak{k}^* \oplus V = \mathfrak{h}^*$. It is not hard to show (cf. [GS6]) that $\mathcal{O}_F$ is symplectomorphic by an H-equivariant map to the H-orbit containing $(i(F), b_0)$.

One then has the following result [S1], [R].

Theorem 2.4.1 *Every coadjoint orbit of H is an "induced" orbit of the form $\mathcal{O}_F$.*

2.5 Riemannian Submersions with Totally Geodesic Fibers

Example 2.8 Let X and B be Riemannian manifolds. A map $\rho: X \to B$ is called a *Riemannian submersion* if it is a submersion and if, for each $x \in X$, the map $d\rho_x: H_x \to TB_{\rho(x)}$ is an isometry. Here H_x denotes the "horizontal subspace at x," that is, H_x is the orthogonal complement relative to the Riemann metric at x of the tangent space to the fiber passing through x: $H_x = (T(\rho^{-1}(y))_x)^\perp$, $y = \rho(x)$. A submanifold W of a Riemannian manifold X is called *totally geodesic* if any geodesic of X tangent to W lies in W.

Suppose that $\rho: X \to B$ is a Riemannian submersion whose fibers are all compact and totally geodesic. Also suppose that X is complete. According to a fundamental theorem of Hermann [Her] there exists a compact Lie group G, a principal G-bundle $P \to B$, a connection A on P, a Riemannian manifold F, and an embedding $G \to$ isometries of F such that

$$
\begin{array}{ccc}
X & \cong & P \times_G F \\
 & \searrow \quad \swarrow & \\
 & B &
\end{array},
$$

such that the vertical component of the metric on X is the metric of F, and such that the vertical–horizontal splitting of the tangent bundle of X is the splitting given by the connection. We can now construct T^*X as a symplectic fiber bundle over T^*B by a combination of the constructions of Examples 2.3 and 2.4: From the data (G, P, B, A, F) we can form the associated bundle

$$P \times_G T^*F$$

as a symplectic fiber bundle over B with a symplectic connection which we shall denote by Γ^D. (Here the superscript D stands for "downstairs.") Now make the base change (Example 2.4) corresponding to the canonical projection $\beta \colon T^*B \to B$. Then we have

$$T^*X \;\cong\; \beta^*(P \times_G T^*F)$$
$$\searrow \qquad \swarrow \qquad ,$$
$$T^*B$$

with connection $\Gamma = \beta^* \Gamma^D$ and corresponding minimal coupling form ω_Γ. Let π denote the projection $\pi \colon T^*X \to T^*B$ and ω_B the canonical two-form on T^*B. Then the canonical two-form $\omega = \omega_X$ on T^*X is given as

$$\omega = \omega_\Gamma + \pi^* \omega_B.$$

Let $g = g_X = \sum g_{ij} dx^i dx^j$ be the Riemannian metric on X, and let g_F and g_B denote its vertical and horizontal components. Let $H \colon T^*X \to \mathbb{R}$ be the Hamiltonian associated to the metric g (the kinetic energy) and $H_B \colon T^*B \to \mathbb{R}$ the Hamiltonian associated to the metric g_B. (So $H(x, \xi) = g(\xi, \xi)$ for $\xi \in T_x^*X$ and $H_B(b, \eta) = g_B(\eta, \eta)$ for $\eta \in T_b^*B$.) Let us write

$$H_F = H - H_B.$$

We can get an alternative description of H_F as follows: Let $H^F \colon T^*F \to \mathbb{R}$ be the kinetic energy associated to the Riemannian metric on F. Since G acts as isometries of F, the function H^F lifts to a function on $P \times_G T^*F$. Let us denote this function by H_{F^D}. Then it follows from the preceding description of T^*X that

$$H_F = \beta^* H_{F^D}.$$

It follows from this description that (1.26) holds, that is, that $\iota(v^\#)dH_F = 0$. Indeed, (1.26) amounts to the assertion that parallel transport with respect to the connection on $P \times_G T^*F$ preserves H_{F^D}, or equivalently, that parallel transport with respect to the connection on $X = P \times_G F$ preserves the Riemannian metric on the fibers. But this is exactly the assertion of the Hermann theorem. We are thus exactly in the situation of Proposition 1.5.2 (except for the change in notation that the base is here T^*B).

Let

$$g_\epsilon = \epsilon^2 g_F + g_B.$$

The limit $\epsilon \to 0$ is known as the adiabatic limit. It has the effect of making distances in B much larger than distances in F. The Hamiltonian H_ϵ corresponding to g_ϵ is then given by (1.27): $H_\epsilon = H_B + \epsilon^2 H_F$. We are thus exactly in the situation described at the end of Section 1.5. Thus equation (1.28) gives the relation between the weak coupling limit and the adiabatic limit.

The two types of spectra we will be concerned with here are $\mathrm{Spec}_Q(X)$, the spectrum of the Laplace operator Δ_X, and $\mathrm{Spec}_C(X)$, the length spectrum of geodesic flow on the cotangent bundle of X. (The latter is defined to be the set of periods of periodic trajectories of the geodesic flow.) If X is compact and $\rho: X \to B$ is a Riemannian submersion with totally geodesic fibers F, then every eigenvalue λ of Δ_X can be written in the form

$$\lambda = \lambda_1 + \lambda_2, \tag{2.22}$$

where λ_1 is an eigenvalue of the so-called horizontal Laplacian on X and λ_2 is an eigenvalue of Δ_F. It is also easy to see that if u is an eigenfunction of Δ_B, $\pi^* u$ is an eigenfunction of Δ_X and the eigenvalues associated with u and $\pi^* u$ are the same. Hence, in particular,

$$\mathrm{Spec}_Q(B) \subseteq \mathrm{Spec}_Q(X). \tag{2.23}$$

If the metric on X is rescaled adiabatically the horizontal Laplacian is unchanged and Δ_F is rescaled by a factor of $\frac{1}{\epsilon^2}$; so, by (2.22), the eigenvalues of the Laplacian on X are now of the form

$$\lambda = \lambda_1 + \frac{\lambda_2}{\epsilon^2}. \tag{2.24}$$

In particular, as $\epsilon \to 0$, eigenvalues of the form $\lambda_1 + \frac{\lambda_2}{\epsilon^2}$, $\lambda_2 \neq 0$, tend to ∞, and the eigenvalues *not* of this form are the eigenvalues in $\mathrm{Spec}_Q(B)$. Thus in particular

$$\mathrm{Spec}_Q(X_\epsilon) \longrightarrow \mathrm{Spec}_Q(B) \tag{2.25}$$

as $\epsilon \to 0$. (See [BBB].)

Our goal in this section is to prove an analog of (2.25) for Spec_C. Our proof will be based upon the interchange of the adiabatic and the weak coupling limit described at the end of the preceding section. We will begin with a few general comments about the objects $\mathrm{Spec}_C(X)$, $\mathrm{Spec}_C(F)$, and $\mathrm{Spec}_C(B)$ and how they are related. There isn't, unfortunately, a classical analog of (2.22);[3] however, as a weak substitute for (2.22), one has:

1. $\mathrm{Spec}_C(F)$ is always contained in $\mathrm{Spec}_C(X)$: Since the fibers of π are totally geodesic, F-geodesics are automatically X-geodesics.
2. Though $\mathrm{Spec}_C(B)$ is not necessarily contained in $\mathrm{Spec}_C(X)$, this does frequently happen. To see why, let γ be a closed geodesic in B of period T_γ. Consider the segment

$$\{\gamma(t) \mid \quad 0 \le t \le T_\gamma\} \tag{2.26}$$

of γ. Let $q = \gamma(0)$ and let p be any point on the fiber above q. The curve (2.26) can be lifted to a curve

$$\{\widetilde{\gamma}(t) \mid \quad 0 \le t \le T_\gamma\} \tag{2.27}$$

in such a way that $\widetilde{\gamma}(0) = p$ and $\frac{d\widetilde{\gamma}}{dt}$ is perpendicular to the fiber at $\widetilde{\gamma}(t)$ for all t. Indeed this lifting is unique; so the map

$$\Phi_\gamma \colon F \longrightarrow F$$

that assigns to p the terminal point on the curve $\widetilde{\gamma}$ is well defined. This map is the *holonomy map* associated with the vertical–horizontal splitting of the tangent bundle of X and, by Hermann [Her], it is an isometry. By O'Neill [O'N], $\widetilde{\gamma}$ is an X-geodesic; hence it is a *closed* X-geodesic iff p is a fixed point of Φ_γ. (Notice, by the way, that if $\widetilde{\gamma}(0) = \widetilde{\gamma}(T_\gamma)$, then $\frac{d}{dt}\widetilde{\gamma}|_{t=0} = \frac{d}{dt}\widetilde{\gamma}|_{t=T_\gamma}$, since both these vectors are horizontal and have the same projection onto B.)

Let G be the group of isometries of F and G_0 the connected component of the identity in G. Suppose B is simply connected. Then γ can be contracted in B to a point, and hence Φ_γ is in G_0. Therefore, if we were able to show that no element of G_0 acts in a fixed-point-free way on F, we would be able to conclude that ϕ_γ has a fixed point and hence that γ has a closed lift to X. In particular if the Euler characteristic $\chi(F)$ is nonzero *every* map that is homotopic to the identity has a fixed point; so we've proved:

---

[3]Except in the case where the sets

$$\left\{ \frac{1}{T^2} \middle| \quad T \in \mathrm{Spec}_C(X) \right\}$$

and

$$\left\{ \frac{1}{T_1^2} + \frac{1}{T_2^2} \middle| \quad T_1 \in \mathrm{Spec}_C(B), \quad T_2 \in \mathrm{Spec}_C(F) \right\}$$

are identical.

Proposition 2.5.1 *If $\pi_1(B) = 0$ and $\chi(F) \neq 0$ then $Spec_C(B) \subseteq Spec_C(X)$.*

Let's now come back to the classical analog of (2.25). We have just seen that, modulo some assumptions about the topology of B and F, $Spec_C(B) \subseteq Spec_C(X_\epsilon)$; so to show that the classical analog of (2.25) is true we are reduced to showing:

Proposition 2.5.2 *Suppose $T_\epsilon \in Spec_C(X_\epsilon)$ tends to a finite limit T as $\epsilon \to 0$. Then T is in $Spec_C(B)$.*

Proof. Since T_ϵ is in $Spec_C(X_\epsilon)$ there exists a periodic trajectory of geodesic flow of period T_ϵ in the unit cosphere bundle of the adiabatically rescaled metric, that is, on the surface

$$H_B + \frac{1}{\epsilon^2} H_F = 1 \tag{2.28}$$

in T^*X. Both H_B and H_F are equal to constants, say $C_{B,\epsilon}$ and $c_{F,\epsilon}$, along this trajectory, and, by (2.28),

$$0 \leq c_{B,\epsilon} \leq 1 \quad \text{and} \quad 0 \leq c_{F,\epsilon} \leq \epsilon^2. \tag{2.29}$$

Applying the homothety $m_{1/\epsilon}$ to this curve we get a periodic curve γ_ϵ of period T_ϵ on the surface

$$H_B = c_{B,\epsilon}, \qquad H_F = \frac{c_{F,\epsilon}}{\epsilon^2}$$

in T^*X. By the results of Section 1.5 this curve is an integral curve of Ξ_ϵ. (See (1.28).) Let p_ϵ be a point on the curve. In view of (2.29) we can assume that as $\epsilon \to 0$, $c_{B,\epsilon}$ and $c_{F,\epsilon}/\epsilon^2$ tend to finite limits, 1 and c, and that p_ϵ tends to a point p on the surface

$$H_F = c, \qquad H_B = 1.$$

Let $\epsilon \omega_\Gamma + \pi^* \omega_B$ be the weak coupling form on T^*X. By definition

$$\iota(\Xi_\epsilon)(\epsilon \omega_\Gamma + \pi^* \omega_B) = dH_X = dH_B + dH_F.$$

Therefore,

$$\Xi_\epsilon = \Xi_{B,\epsilon} + \frac{1}{\epsilon} \Xi_{F,\epsilon}$$

with $\Xi_{B,\epsilon}$ and Ξ_F defined, as in Section 1.5, by

$$\iota(\Xi_{B,\epsilon})(\epsilon \omega_\Gamma + \pi^* \omega_B) = dH_B \tag{2.30}$$

and

$$\iota(\Xi_{F,\epsilon})(\epsilon \omega_\Gamma + \pi^* \omega_B) = dH_F. \tag{2.31}$$

$\square$

The vector field Ξ_F is *vertical* with respect to the symplectic fibration

$$T^*X \longrightarrow T^*B, \qquad (2.32)$$

and equation (2.30) shows that $\Xi_{B,\epsilon}$ tends to a horizontal limit Ξ as $\epsilon \to 0$. Indeed, Ξ is just the horizontal lift to T^*X of the geodesic vector field Ξ^0 on T^*B.

Let γ be the integral curve of Ξ through the point q. Though γ is not necessarily closed we will prove:

Proposition 2.5.3 *The projection of γ onto T^*B is a periodic trajectory of geodesic flow of period T in the unit cosphere bundle of T^*B.*

Proof. Since Ξ is the horizontal lift of the geodesic vector field on T^*B it is clear that the projection of γ is a trajectory of geodesic flow and since $H_B = 1$ on γ it is clear that this trajectory sits in the unit cosphere bundle at B. Therefore what is at issue is showing that the trajectory is periodic of period T. Let $\gamma_\epsilon(t)$, $0 \le t \le T_\epsilon$, be the integral curve of $\Xi_{B,\epsilon}$ through the point p_ϵ. Then as $\epsilon \to 0$ this curve tends to the curve $\gamma(t)$, $0 \le t \le T$. Moreover, since the flows generated by $\Xi_{B,\epsilon}$ and Ξ_F commute (see Section 1) the integral curve of Ξ_ϵ through p_ϵ (that is, the curve $\gamma_\epsilon(t)$, $0 \le t \le T_\epsilon$) is just the curve

$$\exp\left(\frac{1}{\epsilon}t\,\Xi_F\right) \circ \gamma_\epsilon^\flat(t), \qquad 0 \le t \le T_\epsilon,$$

so it has the same projection onto T^*B as $\gamma_\epsilon^\flat(t)$. However, by assumption, $\gamma_\epsilon(t)$ is periodic of period T_ϵ; hence, so is its projection and, therefore, so is the projection of $\gamma_\epsilon^\flat(t)$. However, $\gamma_\epsilon^\flat$ converges to the projection of γ, from which it follows that the projection of γ is periodic of period T. □

2.6 The Hannay–Berry Connection

Example 2.9 Everyone who has taken an elementary course in physics is familiar with the famous experiment of Foucault for demonstrating the rotation of the earth: Suspending a pendulum 61 meters in length from the ceiling of the Pantheon in Paris, Foucault watched the patterns it traced out in a bed of sand. Since the pendulum was oscillating in a single plane these patterns were straight lines; however, the directions of these lines changed slowly in time, reflecting the fact that the direction of the force of gravity was rotating along with the motion of the earth.

One rather surprising feature of Foucault's experiment was that the change in slope of these lines turned out *not* to be a periodic function with one day periods: At the end of one day there was a *phase shift* in the slope. Recently Hannay and Berry have developed a theory to account for phase shifts of this type in dynamical systems in which certain parameters are changing "adiabatically," that is, slowly in time. (In this example this parameter is the direction of gravity.) We won't be able to describe this theory in detail; however, we will give a brief indication of what it has to do with symplectic fibrations. The description we will give here is due to Richard Montgomery. (See [M3]. For other treatments of the material see [Ber], [BH], [Duis], [GKM], [Han], and [MMR].)

Let B be an arbitrary k-dimensional manifold (which, in the discussion to follow, will be the set on which the adiabatic variables that we are concerned with live). Let (X, ω) be a fixed symplectic manifold and G a compact Lie group. Assume that for every $b \in B$ we are given an action τ_b of G on X. We will say that τ_b depends smoothly on b if the mapping

$$\tau: G \times B \times X \to B \times X, \quad (g, b, x) \mapsto (b, \tau_b(g) \cdot x) \tag{2.33}$$

is smooth. Notice that we can think of (2.33) as a rather twisted action of G on the product manifold $B \times X$. Consider now the fibration

$$pr_1: B \times X \to B, \tag{2.34}$$

given by projection onto the first factor. This is obviously a symplectic fibration, and there is a very trivial way of equipping it with a symplectic connection: At (b, x) take the horizontal space for the connection to be $T_b B$. Notice that this connection is defined (in the sense of equation (1.4)) by a closed two-form: the two-form

$$\Omega = (pr_2)^* \omega,$$

where pr_2 is the projection $B \times X \to X$. This connection is, of course, not a terribly interesting connection: The holonomy attached to every closed loop in B is just the identity. We can, however, construct from it a much more interesting connection. Namely, take the two-form Ω and average it with respect to the action (2.33), that is, let

$$\Omega_G = \int_G \tau(g)^* \Omega \, dg, \tag{2.35}$$

dg being Haar measure. This form is closed, and it still has the property that its restriction to each fiber of the fibration (2.34) is ω. Therefore, by the general theory developed in Chapter 1 it gives rise to a G-invariant symplectic connection on the fiber bundle (2.34). (Notice that the trivial connection is *not* G-invariant.)

This connection is Montgomery's version of the Hannay–Berry connection. Let us now see what it has to do with the Foucault pendulum. The *simple spherical pendulum* consists of a plumb bob attached by a steel rod to a fulcrum with a rotating socket (so that it can swivel in all directions) and acted on by the force of gravity (which is assumed to be constant and pointed in a fixed direction, say α). For a description of the dynamics of this system see [Duis].

The configuration space for this system is the sphere S^2 of radius l, l being the length of the steel rod. The phase space X is, therefore, the cotangent bundle of the sphere T^*S^2. The most important feature of this system for us is that it possesses a rotational symmetry about the the the direction of the gravitational field. This is expressed by a Hamiltonian action of S^1 on X that commutes with the dynamical flow. Therefore, since we are dealing with a system of two degrees of freedom, this means that this system is *completely integrable*. In particular, modulo some monodromy problems that we won't concern ourselves with here (see [Duis]) the action of S^1 that we have just described extends to an action of $T^2 = S^1 \times S^1$,

$$\tau_\alpha \colon T^2 \times X \to X,$$

depending, of course, on the direction of gravity α.

Now let $\alpha_0 = (0, 0, -1)$ and, for this choice of the direction of the gravitational field, let γ_0 be a Foucault trajectory of the spherical pendulum, that is, a trajectory lying in a fixed vertical plane. Let p_0 be a point on this trajectory and let Λ_0 be the orbit of T^2 passing through p_0. It is easy to see that Λ_0 is a Lagrangian submanifold of X, and it contains, of course, the trajectory γ_0.

Now let us move α along a closed loop $\mathcal{L}$ in our parameter space $B = S^2$ joining α_0 to itself. (In Foucault's experiment $\mathcal{L}$ is the line of latitude through Paris.) As α moves, the Lagrangian manifold Λ_0 gets carried by parallel transport into a Lagrangian manifold Λ_α, and the trajectory γ_0 into a trajectory γ_α sitting inside Λ_α. Now what happens when α comes back to the point α_0? It is easy to see that Λ_α comes back again to Λ_0. Indeed, Λ_α can be thought of as a fixed level set of the momentum map $\Phi_\alpha \colon X \to \mathfrak{t}^*$ associated with the Hamiltonian action τ_α of T^2, and these level sets are preserved by the parallel transport. (In fact the momentum map Φ_α itself is parallel.) The trajectory γ_0 need *not*, however, come back to itself, but can come back to any other trajectory on Λ_0. Since every trajectory on Λ_0 can be obtained from γ_0 by the rotational action of S^1, this new trajectory has to be a rotated Foucault trajectory, and the angle of rotation is a holonomy invariant of the loop $\mathcal{L}$. In [M3] Montgomery computes this angle and shows that it does, indeed, give the correct Foucault shift in phase.

3

Duistermaat–Heckman Polynomials

3.1 Orbital Methods in Representation Theory

In this section we will discuss (in a rather heuristic spirit, without attempting to be too rigorous) some of the motivating ideas involved in the interaction between symplectic geometry and representation theory.

Let G be a compact, connected Lie group and ρ a finite-dimensional representation of G. The problem we will be concerned with is how to decompose ρ into a sum of irreducible subrepresentations, or, in other words, how to compute the multiplicity $*(\tau, \rho)$ with which each irreducible representation τ of G occurs as a subrepresentation of ρ. For instance, if H is a compact semisimple Lie group, G its Cartan subgroup, and ρ is the restriction to G of an irreducible representation of H, then the irreducible representations of G are one-dimensional representations (corresponding to the weights of G), and our problem comes down to computing the weight multiplicities of ρ. The solution to this problem is provided by the remarkable multiplicity formula of Kostant:

$$*(\alpha, \rho) = \sum (-1)^w N(w(\lambda + \delta) - (\alpha + \delta)).$$

On the right-hand side, the ws are the elements of the Weyl group of H, δ is half the sum of the positive roots, λ is the maximal weight of the representation ρ, $(-1)^w$ is $(-1)^{l(w)}$ where $l(w)$ is the length of w, and N is the Kostant partition function: $N(\upsilon)$ is the number of ways in which one can write the weight υ as a sum of positive roots, that is, the number of solutions, in integers $(n_1, \ldots, n_k)$, of the equation

$$\upsilon = n_1 a_1 + \cdots + n_k a_k \tag{3.1}$$

with $n_1, \ldots, n_k \geq 0$, the a_is being the positive roots. (See, for instance, [J], page 267, or [Hum].) Unfortunately the number of solutions of (3.1) is quite hard to compute if λ is large, and so for large λ various schemes have been devised for

54

estimating $N(v)$. One such scheme is due to Heckman and motivates much of what follows in this chapter. Heckman's scheme consists simply of estimating $N(v)$ by the volume of the polytope

$$\{x_1, \ldots, x_k \geq 0 \mid v = x_1 a_1 + \cdots + x_k a_k\} \qquad (3.2)$$

in $x_1, \ldots, x_k$-space. This volume turns out to be a good deal easier to compute than $N(v)$ itself. For instance, as we will see in Section 3.3, it is a piecewise polynomial function of v.

The "orbital" in our title is a reference to the so-called coadjoint orbit method in representation theory. If H is a nilpotent Lie group, then, according to Kirillov, there is a one-to-one correspondence between the irreducible unitary representations of H and the coadjoint orbits of H; and, with small caveats, the same is true (according to Auslander, Kostant, Duflo, and others) for other types of Lie groups as well. In particular, for the situation that we are considering here – H a compact, connected Lie group – Kostant showed that the irreducible representations of H are in one-to-one correspondence with the *integral* coadjoint orbits of H. (An orbit $\mathcal{O}$ of H is integral if the cohomology class $[\omega_{\mathcal{O}}]$ is an integral element of $H^2(\mathcal{O}, \mathbb{R})$.) This correspondence operates as follows: For every orbit $\mathcal{O}$ there is a unique H-invariant Kähler structure on $\mathcal{O}$ compatible with its symplectic structure; and if $\mathcal{O}$ is integral, its Kähler form is the curvature form of a Hermitian line bundle. The representation associated to $\mathcal{O}$ is the natural representation of H on the holomorphic sections of this line bundle. (This is Kostant's formulation of the Bott–Borel–Weil theorem. See [K1].)

Given this "orbital picture" of the representations of H, Heckman posed in his thesis the following problem. Let ρ be an irreducible representation of H, and $\mathcal{O}$ the coadjoint orbit it corresponds to. Given a closed connected subgroup G of H, can one compute, for each irreducible representation τ of G, the multiplicity $*(\tau, \rho)$ in terms of the geometry of the symplectic manifold $\mathcal{O}$? One can pose this question of Heckman, albeit not quite so precisely, without any reference to the overgroup H in which G sits: Namely, let M be a symplectic manifold on which G acts; that is, let

$$\kappa: G \to \text{Symplecto}(M) \qquad (3.3)$$

be a Hamiltonian action of G on M. Suppose one can "quantize" κ by associating with it in some intrinsic way a unitary representation ρ of G on a finite-dimensional Hilbert space. Given an integral coadjoint orbit $\mathcal{O}$ of G, can one compute the multiplicity with which the irreducible representation corresponding to $\mathcal{O}$ occurs in ρ by "purely geometric means"? In other words – can one express this multiplicity as a symplectic invariant of the pair $(M, \mathcal{O})$?

Example 3.1 Let G, M, ρ, and $\mathcal{O}$ be as before, and let τ be the irreducible representation of G associated with $\mathcal{O}$. Let H_ρ and H_τ be the finite-dimensional Hilbert spaces on which these representations live. Then the multiplicity with which τ occurs in ρ is the dimension of the space of mappings of H_τ into H_ρ that intertwine τ and ρ, that is,

$$\mathrm{Hom}_G(H_\tau, H_\rho). \tag{3.4}$$

The Marsden–Weinstein theorem provides a natural symplectic analog for this space of intertwining maps. Namely, let X be the product symplectic manifold

$$\mathcal{O}^- \times M. \tag{3.5}$$

(The "$-$" superscript on the $\mathcal{O}$ means that the symplectic structure on $\mathcal{O}$ has been changed by replacing the symplectic form $\omega_\mathcal{O}$ by $-\omega_\mathcal{O}$.) Let $\Phi: X \to \mathfrak{g}^*$ be the moment map associated with the action of G on X, and let X_0 be the preimage, with respect to this mapping, of zero (i.e., $X_0 = \Phi^{-1}(0)$). To simplify let's assume that zero is a regular value of Φ and, therefore, that X_0 is a submanifold of X of codimension equal to the dimension of G. Since Φ is G-equivariant, the action of G takes X_0 into itself; and, in fact, the assumption that zero is a regular value of Φ is equivalent to the assumption that the action of G on X_0 is locally free. (See section 24 of [GS8].) For the moment let's simplify further by assuming that the action of G on X_0 is free. Since G is compact, this implies that the orbit space X_0/G is a manifold and that the point–orbit mapping

$$\pi: X_0 \to X_0/G$$

is a principal G-fibration. Henceforth we will denote this orbit space by $M_\mathcal{O}$ and refer to it as the reduced space associated with $\mathcal{O}$. By the Marsden–Weinstein theorem it possesses a unique symplectic form $\omega_{M_\mathcal{O}}$ with the property that

$$\pi^* \omega_{M_\mathcal{O}} = i^* \omega_X,$$

i being the inclusion map of X_0 into X. We will now explain in what sense $M_\mathcal{O}$ is the symplectic analog of the space of intertwining maps (3.4). In symplectic geometry a "map" or "morphism"

$$\Gamma: U \to V$$

from a symplectic manifold U to a symplectic manifold V is a *canonical relation* (in other words, an imbedded Lagrangian submanifold Γ of $U^- \times V$). For instance, if U is the canonical zero-dimensional symplectic manifold consisting of a single point, a "map" Γ of U into V is just a Lagrangian submanifold of V:

The "points" of V in the categorical sense (the morphisms of this "point-object" into V) are the Lagrangian submanifolds of V. Now we claim that $M_{\mathcal{O}}$ has the following property with respect to G-equivariant "maps" of $\mathcal{O}$ into M.

Theorem 3.1.1 *To every "map"* Γ: pt $\to$ $M_{\mathcal{O}}$ *there corresponds a G-equivariant "map"* $\Gamma^{\#}$: $\mathcal{O} \to M$ *(that is, a G-invariant Lagrangian submanifold of $\mathcal{O}^- \times M$). Moreover, if G is semisimple, this correspondence is bijective.*

(See [GS3], Theorem 2.6.) This theorem says that $M_{\mathcal{O}}$ is the symplectic analog of the space (3.4) in the sense that its "points" are the equivariant intertwining "maps" from $\mathcal{O}$ to M. That is, if one "quantizes" $M_{\mathcal{O}}$, the quantum object one gets is the space of all G-equivariant linear mappings from the vector space associated with $\mathcal{O}$ to the vector space associated with M.

In defining $M_{\mathcal{O}}$ we had to make two assumptions about the action of G on X. One was the assumption that the origin in $\mathfrak{g}^*$ is a regular value of the moment map Φ. This assumption does not always hold; however, it does hold for generic choices of $\mathcal{O}$, so it is not very restrictive. The second assumption, that the action of G on X_0 be free, is, unfortunately, much more restrictive. As we have pointed out, if the first assumption holds, the action of G on X_0 is locally free. However, it need not be free even for generic choices of $\mathcal{O}$. (For instance, let G be the simplest of all compact groups: the group S^1. There are lots of examples of symplectic actions of S^1 for which all the orbits are circles, that is, there are no fixed points, but some circular orbits are "shorter" than others. The generic orbits spiral around these exceptional orbits a certain number of times before closing up. In fact this situation can occur *stably*: It can't be eliminated by a small perturbation.) In such cases $M_{\mathcal{O}}$ is no longer a manifold; it is instead what is called a symplectic "orbifold." (See, for instance, [Sa] for a definition of an orbifold called there a "V-manifold.")

If zero is not a regular value of Φ, $M_{\mathcal{O}}$ becomes a more complicated object still. All one can say about such objects at present is that they have the structure of a "symplectic stratified space." (See [SL].)

Example 3.2 The foregoing heuristics lead one to conjecture that the dimension of (3.4) is a symplectic invariant of $M_{\mathcal{O}}$. For instance, suppose that $M_{\mathcal{O}}$ is the empty set. One would then conjecture that (3.4) is just the zero space, that is, that τ doesn't occur as a subrepresentation of ρ. This turns out to be true for the representations that come from coadjoint orbits of a compact group H in which G sits as a subgroup (see [Hec]) and, in fact, for all representations that are constructed, as these representations are, by "Kählerian means." (See [GS3].)

Example 3.3 If $M_{\mathcal{O}}$ is a point, then one's guess would be that (3.4) is one-dimensional, that is, that τ sits inside ρ in one and just one way. For the examples described already this is by and large true (with a few exceptions). Those spaces for which $M_{\mathcal{O}}$ is either a single point or is empty for every coadjoint orbit $\mathcal{O}$ form a very interesting class of manifolds called "multiplicity-free spaces" [GS7] or "Noether integrable" spaces [MF].

Example 3.4 Let ρ be one of the representations considered above; that is, let H be a compact Lie group containing G, M an integral coadjoint orbit of H, and ρ the irreducible representation of H associated with M. Let ρ_n be the irreducible representation of H associated with the coadjoint orbit

$$M_n = \{n\xi \in \mathfrak{h}^* \mid \xi \in M\}$$

and τ_n the irreducible representation of G associated with the coadjoint orbit

$$\mathcal{O}_n = \{n\xi \in \mathfrak{g}^* \mid \xi \in \mathcal{O}\}.$$

Then, for n large, the following is true:

$$* \, (\tau_n, \rho_n) = n^d \mathrm{volume}(M_{\mathcal{O}}) + O(n^{d-1}). \tag{3.6}$$

The first term on the right is the volume of the symplectic manifold that one gets when one reduces M_n by $\mathcal{O}_n$; therefore, (3.6) can be interpreted as saying that the volume of $M_{\mathcal{O}}$ is a very good "first approximation" to the multiplicity with which τ occurs in ρ. (For a proof of (3.6) see [GS8].)

Example 3.4′ The previous paragraph suggests that the symplectic invariant we are looking for should have the following property: Let X be a symplectic manifold, and let X_n be the symplectic manifold X with its symplectic form rescaled by a factor of n. Then, for n large, our invariant (call it β) should satisfy

$$\beta(X_n) \sim \mathrm{volume}(X_n) = n^d \mathrm{volume}(X). \tag{3.7}$$

One intriguing invariant of this type is the Fefferman invariant: the maximum number of unit balls that can be embedded symplectically in X in such a way that there is no overlap. (Gromov has shown that this is a much more delicate symplectic invariant than the volume: In particular, it can distinguish between "long, thin" and "short, squat" symplectic manifolds of the same volume. See [Gr].)

Example 3.5 Let X be a compact symplectic manifold, and let $\mathbb{L}$ be the (infinite-dimensional) manifold whose points are the "points" of X, that is, the embedded Lagrangian submanifolds of X. Weinstein has observed that this manifold has a natural foliation: If W is a Lagrangian submanifold of X and p the point in $\mathbb{L}$ represented by W, the tangent space to $\mathbb{L}$ at p is the space $Z^1(W)$, the space of closed one-forms on W. Contained inside this space is the subspace $B^1(W)$, the space of exact one-forms, and this is the tangent space at p to the leaf of the Weinstein foliation passing through p. Following Weinstein we will call the leaves of this foliation *isodrasts*. Suppose now that the symplectic form on X is integral; that is, suppose that there exists a line bundle-with-connection sitting over X whose curvature form is the symplectic form. If W is a Lagrangian submanifold, the restriction to W of the curvature form (i.e., the symplectic form) vanishes, so the restriction of the line bundle to W is locally flat. W is said to satisfy the *Bohr–Sommerfeld condition* if this restriction is globally flat, that is, admits a nonvanishing autoparallel section. It is not hard to show that if one point on an isodrast satisfies the Bohr–Sommerfeld condition, so do all points on the isodrast; and if this is the case we will refer to the isodrast as a Bohr–Sommerfeld isodrast. For a discussion of this circle of ideas see [W4].

Conjecture 3.1.2 *The number of such isodrasts is a dimensional invariant of the type (3.7).*

At the moment it is not even known if the number of such isodrasts is finite; however, it is known that if one considers instead of $\mathbb{L}$ itself suitable finite-dimensional submanifolds of $\mathbb{L}$, then, in certain instances, the conjecture is true. For a result of this sort involving coadjoint orbits see [GS5].

Example 3.6 From the Hirzebruch–Riemann–Roch theorem one gets another invariant of the type described in the last two examples, the *Riemann–Roch number* of the symplectic manifold X. This is defined to be the integral

$$RR(X) = \int \exp(\omega_X)\tau_X, \qquad (3.8)$$

τ_X being the Todd class of X. We briefly recall the definition of τ_X. (As a good reference, cf. [BGV].) Every compact symplectic manifold X possesses an almost-complex structure $\mathcal{T}$. Moreover, $\mathcal{T}$ is unique up to topological equivalence. (See Steenrod [St], p. 214.) Let $c_i = c_i(\mathcal{T})$ be the ith Chern class of $\mathcal{T}$ ([St], p. 210). From the c_is one constructs the Todd class of X by the following procedure of Hirzebruch. One writes the expression

$$\prod_{i=1}^{n} x_i/(1 - \exp(-x_i))$$

as a function of $w_1, \ldots, w_n$, where w_i is the ith symmetric polynomial in $x_1, \ldots, x_n$. Substituting c_i for w_i in this expression one gets a cohomology class, $\tau \in H^*(X, Q)$. This is by definition the *Todd class* of X. Using the Atiyah–Singer version of the Riemann–Roch theorem one can show that the Riemann–Roch number is equal to the dimension of $\mathrm{Hom}_G(H_\tau, H_\rho)$ in a lot of interesting cases.

Example 3.7 From now on we will assume that G is an r-dimensional torus acting effectively on M. Let $\Phi: M \to \mathfrak{g}^*$ be the moment map associated with the action of G on M and let M_G be the fixed-point set of G in M. The convexity theorem of [At] and [GS2] says that Φ maps M_G onto a finite set and that the entire image of Φ is the convex hull of the points in this set. In particular the image is a convex polytope. Let us denote this polytope by Δ, and let Δ_0 be the subset of Δ consisting of the regular values of Φ. Since M is compact, Δ_0 is an open subset of Δ. We will denote by $\Delta_1, \Delta_2, \ldots$ its connected components. One can show that they are also convex polytopes. Moreover, to each Δ_i there is associated a polynomial function f_i, which we will call the *Duistermaat–Heckman polynomial* (D–H polynomial for short), defined as follows. Let υ be the symplectic measure on M (defined by the wedge product of the symplectic form with itself to the appropriate power), and let $\Phi_* \upsilon$ be its push-forward with respect to Φ. A standard fact about push-forwards says that on each Δ_i the measure $\Phi_* \upsilon$ is absolutely continuous with respect to Lebesgue measure and can in fact be written as the product of Lebesgue measure with a smooth function. This will be, by definition, the function f_i. Duistermaat and Heckman have proved that this smooth function is actually a polynomial, and that

$$(\dim M)/2 - \dim G \geq \text{degree } f_i. \tag{3.9}$$

(This theorem is sometimes called the Archimedes–Duistermaat–Heckman theorem since a special case of it is due to Archimedes: the case $M = S^2$ with the group S^1 acting on S^2 as rotations about the z-axis.) Another way of defining the f_is is as follows. Since G is a torus, the coadjoint orbits are just the points of $\mathfrak{g}^*$. For $\alpha \in \Delta_i$ let M_α be the reduced space associated with (the coadjoint orbit) α. Then

$$f_i(\alpha) = \text{volume}(M_\alpha). \tag{3.10}$$

In other words, the volume of the reduced space M_α varies in a polynomial fashion as α varies in a fixed Δ_i. For a proof of (3.10) see Appendix 3.A to this chapter.

Example 3.8 The way that Duistermaat and Heckman prove this theorem is to show that the symplectic manifold M_α itself varies in a very simple way as α stays inside a fixed Δ_i. Recall that not only is M_α equipped with a symplectic structure, but it has sitting over it a principal G-bundle

$$
\begin{array}{c}
G \;\to\; X_\alpha \\
\downarrow \\
M_\alpha
\end{array}
$$

X_α being the preimage of α (i.e., $X_\alpha = \Phi^{-1}(\alpha)$). It is easy to see that if one forgets about the symplectic structure on M_α, the M_αs are, up to diffeomorphism, the same manifold for all α in a fixed Δ_i, and the G-bundles X_α are, up to G-isomorphism, the same principal bundle. In other words, for each Δ_i one has a differentiable manifold M_i and a principal G-bundle sitting over it

$$
\begin{array}{c}
G \;\to\; X_i \\
\downarrow \\
M_i
\end{array}
\tag{3.11}
$$

and, for each $\alpha \in \Delta_i$, one has a symplectic form ω_α on M_i that varies smoothly as one varies α and becomes the symplectic form on M_α when one identifies M_α with M_i. The result we referred to earlier is the following: Not only does ω_α depend smoothly on α as long as α stays inside Δ_i, but its cohomology class

$$
[\omega_\alpha] \in H^2(M_i, \mathbb{R})
$$

depends *linearly* on α. In fact for α and β in Δ_i

$$
[\omega_\beta] = [\omega_\alpha] + (\beta - \alpha, c),
\tag{3.12}
$$

where c is the Chern class of the principal bundle (3.11). (Notice, by the way, that since G is an abelian group, its Chern class sits in $H^2(M_i, \mathfrak{g})$. Therefore, since $\beta - \alpha$ is in $\mathfrak{g}^*$ the pairing in the second term on the right makes sense. For a proof of (3.12) see [DH1].) Let s be half the dimension of M_i, that is,

$$
2s = \dim M - 2 \dim G.
$$

Then the volume of M_α is just the expression

$$
[\omega_\alpha]^s (M_\alpha).
\tag{3.13}
$$

By (3.12) this expression is a polynomial in α of degree less than or equal to s; and, in view of (3.10), this proves that so are the f_is.

Example 3.9 The Todd class of M_α is also independent of α as long as α stays inside a fixed Δ_i; so the same argument shows that the Riemann–Roch number of M_α is a polynomial of degree less than or equal to s as a function of α as long as α stays inside a fixed Δ_i. We will denote this polynomial by g_i. Notice that g_i and f_i have the same leading term, that is, their difference is of degree strictly less than s.

Remark. Let α be a weight (i.e., an integer lattice point) in Δ_i. We pointed out in Example 3.6 that $g_i(\alpha)$ is, under suitable hypotheses, equal on the nose to the multiplicity of the weight α, or, in other words, that g_i is the multiplicity function associated with Δ_i. This shows that the multiplicity function admits an interpolating polynomial; there is a polynomial (of degree less than or equal to s) whose restriction to the lattice points in Δ_i is the multiplicity function.

Example 3.10 We will make a few additional comments about the formula (3.6). If G is an r-torus, the irreducible representations of G are indexed by the weights of G. Now, for each weight α let $N_{\alpha,k}$ be the multiplicity with which α occurs in the representation ρ_k. A convenient way of encoding the data contained in the multiplicity diagram for ρ_k is by means of the discrete probability measure:

$$\upsilon_k = N_k^{-1} \sum N_{\alpha,k} \delta_{\alpha/k},$$

where $N_k = \sum N_{\alpha,k}$ and δ_υ is the Dirac delta function at υ. The following theorem gives a very precise interpretation of the Duistermaat–Heckman measure $\Phi_*\upsilon$ as an "asymptotic description" of the multiplicity diagram of ρ_k as k tends to infinity.

Theorem 3.1.3 *As k tends to infinity, υ_k tends weakly to the measure $\Phi_*\upsilon$.*

For a proof of this theorem, see [GS8]. For the coadjoint orbit case, a more "hands-on" proof can be found in Heckman's thesis [Hec]. One can also give a sharp estimate for the rate at which υ_k converges.

Example 3.11 Here is a simple (but enlightening) example: Let M be a six-dimensional coadjoint orbit of $SU(3)$, let G be the Cartan subgroup of $SU(3)$ (the group of three-by-three diagonal matrices with the entries $e^{i\theta_1}$, $e^{i\theta_2}$, and $e^{i\theta_3}$ along the diagonal subject to the condition that $\theta_1 + \theta_2 + \theta_3 = 0 \mod 2\pi$), and let G act in the standard way on M (by matrix conjugation). Then the image of the moment map is a hexagonal region of the type depicted in Figure 3.1. For most of the six-dimensional orbits of $SU(3)$, the Δ_is are seven in number and are as depicted in Figure 3.2. There is, however, a one-parameter family of six-dimensional orbits for which the image of the moment map is a regular

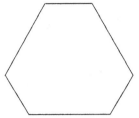

Fig. 3.1.

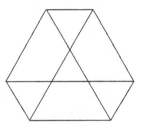

Fig. 3.2.

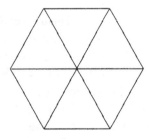

Fig. 3.3.

hexagon, and for these orbits, the Δ_is are six in number and look like the six regions in Figure 3.3. By the Duistermaat–Heckman theorem, the f_is have to be either constant or linear. The f_i corresponding to the interior region in Figure 3.2 is constant and the other six f_is are linear. Their level sets are depicted in Figure 3.4. Each of these f_is is zero on the outer edge of Δ_i and assumes its maximum on the edge next to the interior region. Hence the maximum overall is taken on the interior region. For every Δ and $\alpha \in \Delta_i$, the reduced space M_α is a two-sphere; however, the fibration

$$G \rightarrow X_\alpha$$
$$\downarrow$$
$$M_\alpha$$

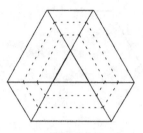

Fig. 3.4.

changes from region to region. For the interior region it is just the trivial fibration

$$S^2 \times G$$
$$\downarrow$$
$$S^2$$

and for each of the six exterior regions it involves the Hopf fibration $\pi: S^3 \rightarrow S^2$. More explicitly, composing π with the projection $S^3 \times S^1 \rightarrow S^3$, one gets a fibration

$$S^3 \times S^1 \rightarrow S^2 \qquad (3.14)$$

which makes $S^3 \times S^1$ into a principal T^2-bundle over S^2. The six elements of the Weyl group provide six distinct ways of identifying G with T^2, and hence six distinct ways of making (3.14) into a principal G-bundle; and these turn out to be the fibrations associated with the six exterior regions of Figure 3.2.

Suppose finally that M is an integral coadjoint orbit of $SU(3)$. Let ρ be the irreducible representation of $SU(3)$ associated with M. Since the Todd class of S^2 is one, the g_is of Example 3.9 are identical with the f_is; so the multiplicities of the weights of ρ are distributed as in Figure 3.5. The dots are the lattice

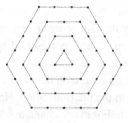

Fig. 3.5.

points (weights). Lattice points lying along a given parallel line have the same multiplicity, and the lattice points in the interior region are the points of highest multiplicity and are all of the same multiplicity. (See [Zel], page 218.)

3.2 Computing the D–H Polynomials: The Case of Linear Action

The rest of this chapter will consist of recipes for computing the f_is. We first show how to compute the f_is on regions Δ_i adjacent to the exterior vertices of Δ. For such regions the action of G can be assumed to be a linear action (see Section 3.3); so we will start by investigating in detail the linear case. To fix our conventions and normalization, let us begin with the case of a circle

$$G = \mathbb{R}/2\pi \mathbb{Z}$$

acting on $\mathbb{C}$, where $\tau \in G$ acts as

$$z \rightsquigarrow e^{\alpha i \tau} z, \qquad \alpha \in \mathbb{Z}.$$

Let us choose our symplectic form to be

$$\omega = i dz \wedge d\bar{z} = 2\, dx \wedge dy = 2r dr \wedge d\theta.$$

(Our choice of the factor of 2 here is to avoid factors of $\frac{1}{2}$ later on.) The infinitesimal generator of our action is the vector field $\alpha \frac{\partial}{\partial \theta}$. We have

$$\iota\left(\alpha \frac{\partial}{\partial \theta}\right)\omega = -2\alpha r dr = -\alpha d(r^2) = -\alpha d|z|^2.$$

Thus the moment map is given by

$$\Phi_\alpha(z) = \alpha|z|^2$$

and the image of the moment map is the half-line

$$\{s\alpha \mid \quad s \geq 0\}.$$

(Here we have identified the Lie algebra of G and its dual space as well with the real line $\mathbb{R}$.)

On any symplectic manifold of dimension $2N$, the form ω^N determines an orientation, and hence an identification between $2N$ forms and densities. If ρ is a $2N$-form, we denote the corresponding density by $|\rho|$. We define the Liouville measure to be $\frac{1}{(2\pi)^N N!}|\omega^N|$.

In the case at hand we have $N = 1$ and

$$\frac{1}{2\pi}|\omega| = \frac{1}{\pi}\, dx dy = \frac{1}{\pi} r dr d\theta.$$

If $\alpha \neq 0$, the push-forward of $\frac{1}{2\pi}|\omega|$ under Φ_α is defined to be the measure on $\mathbb{R}$ given by

$$\frac{1}{2\pi}\langle \Phi_{\alpha*}(|\omega|), f\rangle = \frac{1}{2\pi}\langle|\omega|, f \circ \Phi_\alpha\rangle$$

for any continuous function f of compact support on $\mathbb{R}$. (Here $\langle\,,\,\rangle$ denotes the pairing between measures and functions.) This definition makes sense since the map Φ_α is proper if $\alpha \neq 0$. Explicitly, the right-hand side of the previous equation is

$$\frac{1}{2\pi}\iint_C f(\alpha r^2) 2r\,dr\,d\theta = \int_0^\infty f(\alpha r^2)\,d(r^2)$$
$$= \int_0^\infty f(\alpha s)\,ds$$

if we set $s = r^2$. We can write this as follows. Let dt_+ denote the measure on the one-dimensional space $\mathbb{R}$ that is supported on the positive half-line $0 \leq t < \infty$ and coincides with Lebesgue measure dt there. Let $\iota_\alpha \colon \mathbb{R} \longrightarrow \mathfrak{g}^*$ (identified here with $\mathbb{R}$) be defined by

$$\iota_\alpha \colon \mathbb{R} \longrightarrow \mathfrak{g}^*, \quad \iota_\alpha(t) = t\alpha. \tag{3.15}$$

Then we have proved that

$$\Phi_{\alpha *}\left(\frac{1}{2\pi}|\omega|\right) = \iota_{\alpha *}(dt_+). \tag{3.16}$$

More generally, suppose that G is an n-dimensional torus,

$$G = \mathbb{R}^n / 2\pi \mathbb{Z}^n.$$

Suppose that G acts on $\mathbb{C}$ by $\tau \in G$ acting as

$$z \rightsquigarrow e^{i\alpha \cdot \tau} z.$$

Here α is a weight, that is, $\alpha \in \mathbb{Z}^n \subset \mathbb{R}^n \sim \mathfrak{g}^*$, and $\alpha \cdot \tau$ denotes the value of α on τ. Then all the preceding formulas hold, except that now $\alpha \in \mathbb{Z}^n$ is a vector. Thus the image of the moment map is the ray $\{s\alpha \mid s \geq 0\}$ in the n-dimensional space $\mathfrak{g}^*$, that is, the image of $\mathbb{R}^+$ under the map ι_α defined by (3.15). The push-forward of the Liouville measure is then given by (3.16).

Now let (V, ω) be a symplectic vector space and ρ a symplectic representation of $G = \mathbb{R}^n / 2\pi \mathbb{Z}^n$ on V. We can decompose V into a direct sum of "real weight spaces"

$$V = \oplus W^\alpha, \quad \alpha \in \mathbb{Z}^n \subset \mathbb{R}^n \sim \mathfrak{g}^*. \tag{3.17}$$

Here each W^β denotes the subspace on which the element τ of G acts as a rotation $R_{\beta \cdot \tau}$ through angle $\beta \cdot \tau$. Thus, if $\beta \neq 0$, W^β is a direct sum of two dimensional spaces on each of which G acts as the matrix

$$\begin{pmatrix} \cos \beta \cdot \tau & -\sin \beta \cdot \tau \\ \sin \beta \cdot \tau & \cos \beta \cdot \tau \end{pmatrix}$$

relative to an appropriate basis. On the space W^0 the representation of G is trivial. It follows from the invariance of ω that the spaces W^β and W^α are orthogonal relative to ω if $\beta \neq \alpha$. Indeed, if $\beta \neq \alpha$ then one can find a $\tau \in G$ such that $R_{\alpha \cdot \tau} = \mathrm{Id}$ and $R_{\beta \cdot \tau} - \mathrm{Id}$ is invertible, or vice versa. But if $u \in W^\alpha$ and $v \in W^\beta$ then $\omega(u, v) = \omega(\rho(\tau)u, \rho(\tau)v)$ implies that

$$\omega(u, R_{\beta \cdot \tau} v - v) = 0,$$

or, since $R_{\beta \cdot \tau} - \mathrm{Id}$ is invertible, that $\omega(u, w) = 0$ for all $u \in W^\alpha$ and $w \in W^\beta$. Therefore, each of the spaces W^β is symplectic. We claim that each of the spaces W^β can be decomposed into a direct sum of two-dimensional invariant symplectic subspaces mutually orthogonal with respect to ω. For the case $\beta = 0$ this is obvious. If $\beta \neq 0$ then the rotation R_t through angle t is well defined in W^β and we can define $J = \frac{d}{dt} R_t |_{t=0}$, so

$$J^2 = -\mathrm{Id} \qquad \text{and} \qquad \rho(\tau) = \cos \beta \cdot \tau \mathrm{Id} + \sin \beta \cdot \tau J$$

on W^β. It is enough to show that there is some vector $v \in W^\beta$ such that $\omega(v, Jv) \neq 0$, for the space spanned by v and Jv is invariant and symplectic, and so we can consider its orthogonal complement with respect to ω and proceed by induction. Now the invariance of ω implies that

$$\omega(u, Jv) + \omega(Ju, v) = 0 \qquad\qquad \forall u, v.$$

If $\omega(w, Jw) \equiv 0$ then differentiating $\omega(u + \epsilon v, Ju + \epsilon Jv) = 0$ with respect to ϵ at $\epsilon = 0$ implies

$$\omega(u, Jv) + \omega(v, Ju) = 0.$$

Adding these equations gives $\omega(u, Jv) \equiv 0$, which is impossible since $J^2 = -\mathrm{Id}$. We thus have the orthogonal decomposition

$$V = \oplus V^i \qquad\qquad (3.18)$$

into two-dimensional invariant symplectic subspaces. Now each carries an invariant positive-definite scalar product (determined up to positive multiple if $\beta \neq 0$) and a symplectic form

$$\omega_j = \omega_{|_{V^j}}.$$

Thus V^j carries a complex structure and, in fact, a preferred complex coordinate z_j (determined up to a phase factor) satisfying

$$\omega_j = i \, dz_j \wedge d\bar{z}_j.$$

The action of G on V^j is given by

$$z \rightsquigarrow e^{i\alpha_j \cdot \tau} z, \qquad\qquad \alpha_j \in \mathbb{Z}^n \subset \mathbb{R}^n = \mathfrak{g}^*.$$

If $V^j \subset W^\beta$ then $\alpha_j = \pm\beta$; the form ω_j determines which is the correct sign. Since $\omega = \omega_1 \oplus \cdots \oplus \omega_N$ the moment map for the action of G on V is just the sum of the moment maps for the individual summands, whence

$$\Phi(z_1, \ldots, z_N) = |z_1|^2 \alpha_1 + |z_2|^2 \alpha_2 + \cdots + |z_N|^2 \alpha_N. \qquad (3.19)$$

Thus the image of the moment map is the convex conic polytope

$$\Delta = \{ s_1 \alpha_1 + \cdots + s_N \alpha_N \mid s_1, \cdots, s_N \geq 0 \}. \qquad (3.20)$$

Furthermore, the Liouville measure is the product measure

$$\frac{1}{(2\pi)^N N!} |\omega^N| = \frac{1}{(2\pi)^N} |\omega_1 \wedge \cdots \wedge \omega_N| = \frac{1}{2\pi} |\omega_1| \times \cdots \times \frac{1}{2\pi} |\omega_N|. \quad (3.21)$$

To summarize, we have proved:

Lemma 3.2.1 *Given a linear symplectic action of $G = \mathbb{R}^n/2\pi\mathbb{Z}^n$ on (V, ω) we have the decomposition $V = \oplus V^i$, where each V^i is a two-dimensional symplectic vector space that has the structure of a one-dimensional complex vector space, which is a weight space with weight α_i. The moment map is given by (3.19), its image is (3.20), and the Liouville measure is the product measure (3.21).*

To push forward the Liouville measure we must know if Φ is proper. The following is easy to check:

Lemma 3.2.2 *Φ is proper if and only if Δ is properly contained in a half-space, that is, if and only if there exists a $\xi \in g$ such that $\langle \alpha_i, \xi \rangle > 0$ for all i.*

Let us assume from now on that this is the case. Then $\upsilon_\Phi = \Phi_*(\frac{1}{(2\pi)^N N!} |\omega^N|)$ is well defined. Since $\frac{1}{(2\pi)^N N!} |\omega^N|$ is a product measure, (3.21), and since Φ is the sum of the Φ_i, it follows that υ_Φ is the convolution of the push-forwards of the measures for each V^i. By (3.16) this implies

$$\upsilon_\Phi = \Phi_* \left(\frac{1}{(2\pi)^N N!} |\omega|^N \right) = (\iota_{\alpha_1 *} dt_+) * \cdots * (\iota_{\alpha_N *} dt_+). \qquad (3.22)$$

As we shall see, it is not difficult to compute the right-hand side of (3.22) by induction on N. Another useful formula for the measure υ_Φ is the following. Consider Φ as the composition of the mapping

$$\Psi: V \to \mathbb{R}^N, \quad (z_1, \ldots, z_N) \mapsto (|z_1|^2, \ldots, |z_N|^2)$$

and the linear mapping

$$L: \mathbb{R}^N \to (\mathbb{R}^n)^*, \quad (s_1, \ldots, s_N) \mapsto s_1 \alpha_1 + \cdots + s_N \alpha_N.$$

Lemma 3.2.3 *Let χ be the characteristic function of the positive orthant* $\{s_1 \geq 0, \ldots, s_N \geq 0\}$ *in* $\mathbb{R}^N$. *Then*

$$\Psi_*(|\omega|^N) = (2\pi)^N N! \chi |ds_1 \wedge \cdots \wedge ds_N|.$$

Proof. This is just formula (3.22) with $N = n$ and $\alpha_1, \ldots, \alpha_N$ the standard basis vectors in $\mathbb{R}^N$.   □

Thus we obtain for υ_Φ the formula

$$\upsilon_\Phi = L_*(\chi |ds_1 \wedge \cdots \wedge ds_N|). \tag{3.23}$$

From the right-hand side of this formula one immediately deduces:

Proposition 3.2.4 *If the α_is span* $(\mathbb{R}^n)^*$, *then υ_Φ is absolutely continuous with respect to Lebesgue measure, that is, it can be written in the form*

$$\upsilon_\Phi = f(x)\, dx, \tag{3.24}$$

$f(x)$ *being a locally $\mathcal{L}^1$ function.*

Proof. If the α_is span $(\mathbb{R}^n)^*$, one can make a linear change of coordinates in $\mathbb{R}^N$ so that L becomes the mapping $x_i = s_i$, $i = 1, \ldots, n$; and in these coordinates the proposition is just a consequence of Fubini's theorem.   □

The value of f at any point α in $(\mathbb{R}^n)^*$ is just the volume of the intersection of $L^{-1}(\alpha)$ with the positive orthant: That is, $f(\alpha)$ *is equal to the volume of the set*

$$\{s_1, \ldots, s_N \geq 0 \mid \alpha = s_1\alpha_1 + \cdots + s_N\alpha_N\}. \tag{3.25}$$

Notice that if α and the α_is are large, and α is an integer lattice point, this volume is approximately equal to the number of integer lattice points lying in this set, that is, $f(\alpha) \approx N(\alpha)$, where N is *the partition function* associated with the weights $\alpha_1, \ldots, \alpha_N$. The third description of υ_Φ is as the fundamental solution of a partial differential equation. Suppose $\alpha_i = (a_{i1}, \ldots, a_{in})$. Let

$$D_{\alpha_i} = a_{i1}\frac{\partial}{\partial x_1} + \cdots + a_{in}\frac{\partial}{\partial x_n},$$

and let δ_0 be the Dirac delta function with support at the origin in $(\mathbb{R}^n)^*$. We will show that

$$D_{\alpha_1} \cdots D_{\alpha_N} \upsilon_\Phi = \delta_0. \tag{3.26}$$

Proof. Suppose first that $N = 1$ and $\alpha_1 = \alpha$. Evaluating the left-hand side on a smooth, compactly supported function f, we get

$$\begin{aligned}
(D_\alpha \upsilon_\Phi, f) &= -(\upsilon_\Phi, D_\alpha f) \\
&= -((\iota_\alpha)_* dt_+, D_\alpha f) \\
&= -\int_0^\infty \sum_{i=1}^n a_i \frac{\partial f}{\partial x_i}(a_1 t, \dots, a_n t)\, dt \\
&= -\int_0^\infty \frac{d}{dt} f(a_1 t, \dots, a_n t)\, dt \\
&= f(0),
\end{aligned}$$

which is just δ_0 evaluated on f. To prove (3.26) in general we go back to the formula (3.22) and use the fact that $D(\mu_1 * \mu_2) = D\mu_1 * \mu_2 = \mu_1 * D\mu_2$ for any constant-coefficient operator D. This gives us for the left-hand side of (3.26) the expression

$$(D_{\alpha_1}(\iota_{\alpha_1})_* dt_+) * \cdots * (D_{\alpha_N}(\iota_{\alpha_N})_* dt_+).$$

As we've just seen, each factor in this convolution product is a delta function supported at the origin; hence so is the product itself.   □

By assumption the cone (3.20) is properly contained in some half-space

$$(H_\xi)^+ = \{x \in (\mathbb{R}^n)^* \mid \langle \xi, x \rangle > 0\}.$$

Hence the measure υ_Φ is properly supported in this half-space. However, by a standard theorem on the uniqueness of fundamental solutions of constant-coefficient hyperbolic partial differential equations, there is a *unique* solution of the equation

$$D_{\alpha_1} \cdots D_{\alpha_N} e^+ = \delta_0$$

supported in the half-space $(H_\xi)^+$; and, therefore, this solution has to be υ_Φ. Here is a proof of this assertion in our case. First we show:

Lemma 3.2.5 *Given υ in $\mathbb{R}^n$ and ℓ in $(\mathbb{R}^n)^*$ with $\langle \ell, \upsilon \rangle \neq 0$ let ϕ be a distribution on $\mathbb{R}^n$ that satisfies*

(i) $D_\upsilon \phi = 0$,

(ii) $\phi \equiv 0$ on the half-space $\ell(x) < 0$.

Then $\phi = 0$ everywhere.

Proof. One can take $\ell = x_n$ and $v = (0, \dots, 1)$. This lemma is an easy consequence of the following fact.

Lemma 3.2.6 *Given any function f in $C_0^\infty(\mathbb{R}^n)$, there exist g and h in $C_0^\infty(\mathbb{R}^n)$ such that h is supported in the half-space $\{x_n < 0\}$ and $f = \partial g/\partial x_n + h$.*

Our desired uniqueness result now follows by induction.   □

Notice by the way that if we differentiate v_Φ by just one of the D_{α_i}s, say D_{α_1}, we get, by the same argument as before,

$$D_{\alpha_1} v_\Phi = (\iota_{\alpha_2})_* dt_+ * \cdots * (\iota_{\alpha_N})_* dt_+. \tag{3.27}$$

We will make use of this identity in what follows.

Next let's investigate some properties of the measure v_Φ. For every subset S of the set of weights $\{\alpha_1, \dots, \alpha_N\}$, let

$$(V_0)^S = \prod_{\alpha \in S} (V^\alpha - \{0\}).$$

It is easy to see that at every point p of $(V_0)^S$, the subgroup of G that stabilizes p is independent of p and is the group

$$\{x \in \mathbb{R}^n / 2\pi \mathbb{Z}^n \mid e^{i\langle \alpha, x \rangle} = 1 \ \forall \alpha \in S\}. \tag{3.28}$$

Since a point of V is a critical point of Φ iff its stabilizer group is not discrete, we conclude:

Lemma 3.2.7 *The set of critical points of Φ is a disjoint union of $(V_0)^S$s; moreover, a $(V_0)^S$ is critical iff the αs in S are not a set of spanning vectors of $(\mathbb{R}^n)^*$.*

Let W^S be the following subset of $(\mathbb{R}^n)^*$:

$$W^S = \left\{ \sum s_\alpha \alpha \, \middle| \, \alpha \in S, \ s_\alpha \geq 0 \right\}.$$

From the lemma and the definition of Φ, we get:

Proposition 3.2.8 *The critical values of Φ are the union of the W^Ss for which S is not a spanning set of $(\mathbb{R}^n)^*$.*

Let Δ_0 be the complement in Δ of the set of critical values of Φ, and let Δ_i, $i = 1, \dots, r$, be the connected components of Δ_0. By Proposition 3.2.8, the Δ_is are open conic polytopes, and the sets W^S are the walls of these polytopes. Now let's write v_Φ as the product of Lebesgue measure with a locally $\mathcal{L}^1$-summable function f as in (3.24). Since each Δ_i is contained in the set of regular values of Φ, the restriction of f to Δ_i is a smooth function. Notice also that, by (3.23), the measure υ_Φ is the push-forward by a linear map of a measure on $\mathbb{R}^N$ which is homogeneous of degree N with respect to the group of homotheties of $\mathbb{R}^N$. Thus υ_Φ is also homogeneous of degree N, and so, by (3.23), f is homogeneous of degree $N - n$, that is,

$$f(tx) = t^{N-n} f(x).$$

We will now prove the Duistermaat–Heckman theorem in this linear setting:

Theorem 3.2.9 *The restriction of f to each Δ_i is a homogeneous polynomial of degree $N - n$.*

Proof. Choose coordinates in $(\mathbb{R}^n)^*$ so that Δ is properly contained in the half-space $x_1 \geq 0$ and that α_1 is the unit vector pointing in the direction of the positive x_1-axis. By (3.27)

$$\frac{\partial}{\partial x_1} f = (\iota_{\alpha_2})_* dt_+ * \cdots * (\iota_{\alpha_N})_* dt_+. \tag{3.29}$$

Suppose, to begin with, that $n = N$. Then $\alpha_1, \ldots, \alpha_N$ are a basis of $(\mathbb{R}^n)$ and the assertion is clear. Now assume by induction that the right-hand side is a sum of the form

$$\sum g_i(x_1, \ldots, x_n) \chi_i,$$

where the g_is are polynomials and the χ_is are the characteristic functions of the Δ_is. Let p be a generic point in Δ_0 and let $x_1, \ldots, x_n$ be its coordinates. By integrating (3.29), we get

$$f(x_1, \ldots, x_n) = \sum \int_{-\infty}^{x_1} g_i(s, x_2, \ldots, x_n) \chi_i ds. \tag{3.30}$$

For a generic point p in Δ_0, the ray $p + t\alpha_1$, $-\infty < t < 0$, intersects the $(n - 1)$-dimensional walls of the Δ_is transversally and doesn't intersect any of the lower-dimensional walls. Thus, since each Δ_i is convex, one of the three following alternatives has to be true:

1. The ray doesn't intersect the boundary of Δ_i at all.
2. It intersects the boundary of Δ_i in just one point (in which case p is an interior point of Δ_i).
3. It intersects the boundary of Δ_i in two points.

Moreover, in the last two cases, the points of intersection depend linearly on p: That is, in the second case the point of intersection

$$p' = (x', x_2, \ldots, x_n)$$

satisfies a linear equation

$$x' = a_1 x_1 + \cdots + a_n x_n,$$

and in the third case the points of intersection

$$p' = (x', x_2, \ldots, x_n)$$

and

$$p'' = (x'', x_2, \ldots, x_n)$$

satisfy linear equations

$$x' = a_1 x_1 + \cdots + a_n x_n$$

and

$$x'' = b_1 x_1 + \cdots + b_n x_n.$$

In the first case the ith term makes no contribution at all to the sum (3.30). In the second case it makes the contribution

$$\int_{x'}^{x_1} g_i(s, x_2, \ldots, x_n) \, ds,$$

and in the third case, the contribution

$$\int_{x'}^{x''} g_i(s, x_2, \ldots, x_n) \, ds.$$

It is clear in either case that this expression is a polynomial function of the x_is.
$\qquad\qquad\qquad\qquad\qquad\qquad\qquad\qquad\qquad\qquad\qquad\qquad\qquad\qquad$ $\square$

Remark. This proof can be converted into a fairly efficient algorithm for computing f.

Next we will derive a formula for the "jumps" in f across walls separating two adjacent Δ_is. (We will take pains, by the way, to write this formula as "intrinsically" as possible, because, as we will see in Section 5, the version of this formula that we will give is true in the manifold setting as well.) Let $W = W^S$ be an $(n-1)$-dimensional wall separating the regions Δ_+ and Δ_-, and let G^S be the subgroup of G defined by the set of equations (3.28). Since the $\alpha \in S$ span an $(n-1)$-dimensional subspace of $(\mathbb{R}^n)^*$, this group is of dimension one. Let ξ be a nonzero element in its Lie algebra. It is clear that

$\langle \xi, \alpha \rangle = 0$ for all $\alpha \in S$, and that these equations determine ξ up to a constant multiple. Conversely we can assume that S consists exactly of those weights for which $\langle \xi, \alpha \rangle = 0$. We will fix the orientation of ξ by requiring that it be the outward normal to the region Δ_-. With this convention ξ is determined up to a positive constant multiple.

Let V^S be the subspace of V spanned by the W^αs in the sum (3.17) with $\alpha \in S$. By (3.28), V^S is the fixed-point set of the group G^S; so, by restricting ρ to V^S, we get a representation of the quotient group G/G^S on V^S. Its moment map is just the restriction of Φ to V^S and maps V^S onto the $(n-1)$-dimensional wall W^S. We will denote by υ_S the analog of the measure υ_Φ for the action of G/G^S on V^S and think of this measure as living on W^S. We will show that this measure is all the data needed to compute the jump in f across W^S.

Just as for υ_Φ we can write υ_S as the product of a locally $\mathcal{L}^1$ function f_S (defined on W_S) times the Lebesgue measure on W^S. A slight hitch is that the Lebesgue measure on W^S is only defined up to multiplication by a positive constant. However, the choice of ξ gives us a way of fixing this constant. Let $(\xi_1, \ldots, \xi_n)$ be the coordinates of ξ and let ν_S be an $(n-1)$-form on $(\mathbb{R}^n)^*$ of the form

$$\nu_S = \sum_{i=1}^n a_i (-1)^i \, dx_1 \wedge \cdots \wedge \widehat{dx_i} \wedge \cdots \wedge dx_n,$$

the a_is being constants which satisfy

$$\sum a_i \xi_i = 1. \tag{3.31}$$

Then the restriction of ν_S to W^S doesn't depend on the choices of the a_is and defines both an orientation and a measure on W^S. This measure, which we will continue to denote by ν_S, will be by definition our Lebesgue measure on W^S. In terms of it we can write

$$\upsilon_S = f_S \nu_S. \tag{3.32}$$

Notice, by the way, that f_S is itself (by Theorem 3.2.9 applied to the action of G/G^S on V^S) a piecewise-polynomial function. Also notice that if we multiply ξ by a positive number λ then by (3.31) ν_S gets multiplied by a factor of $1/\lambda$ and hence f_S gets multiplied by a factor of λ.

Before stating our main result we need two final pieces of notation. Via the identification $\mathbb{R}^n \cong (\mathbb{R}^n)^{**}$ we can think of ξ as a linear functional on $(\mathbb{R}^n)^*$. It will cause untold confusion later if we use the same notation for ξ and for this linear functional, so we will denote this linear functional by L_ξ. A second bit of notation that we will need is the following. In the decomposition (3.17) we can assume that the α_is are so labeled that the first m α_is are not in S and the remaining α_is are.

Theorem 3.2.10 *Suppose f is equal to the polynomial f_+ on Δ_+ and f_- on Δ_-. Then*

$$f_+ - f_- = \left(\frac{1}{(m-1)!} \prod_{i=1}^{m} \langle \alpha_i, \xi \rangle^{-1} \right) f_S L_\xi^{m-1} + g, \qquad (3.33)$$

g being a polynomial that vanishes to order m on W^S.

Proof (by induction on m). Assume by induction that along W^S the distribution

$$(\iota_{\alpha_2})_* dt_+ * \cdots * (\iota_{\alpha_N})_* dt_+$$

has a singularity of the form

$$\left(\frac{1}{(m-2)!} \prod_{i=2}^{m} \langle \alpha_i, \xi \rangle^{-1} \right) f_S L_\xi^{m-2}.$$

By (3.27) this distribution is the derivative of f with respect to D_{α_1}. However,

$$L_\xi^{m-2} = \frac{\langle \alpha_1, \xi \rangle^{-1}}{m-1} D_{\alpha_1} L_\xi^{m-1},$$

so f itself has to have a singularity of the form (3.33) along W^S.   □

Remark. Notice that the number of weights in S is at least $n-1$; so m is less than or equal to $N-n+1$. The formula (3.33) is particularly simple when this inequality is an equality, that is, when $m = N - n + 1$. Then, since f_+ and f_- are homogeneous polynomials of degree $N - n$, g has to be zero and f_S has to be a constant. This constant is easy to compute: Let

$$\alpha_{m+1} = (a_{1,1}, \ldots, a_{1,n}),$$
$$\vdots \qquad\qquad\qquad (3.34)$$
$$\alpha_N = (a_{n-1,1}, \ldots, a_{n-1,n}),$$

and let A_ξ be the matrix having the vectors (3.34) as its first $n-1$ rows and $(a_1, \ldots, a_n)$ as its last row (the a_is being as in (3.31)). Then

$$f_S = c_S = \det(A)^{-1} \qquad (3.35)$$

and (3.33) reduces to

$$f_+ - f_- = \left(\frac{c_S}{(N-n)!} \prod_{i=1}^{N-n+1} \langle \alpha_i, \xi \rangle^{-1} \right) (L_\xi)^{N-n}; \qquad (3.36)$$

that is, the jump across W^S is a *constant* multiple of the monomial $(L_\xi)^{N-n}$.

Example Let $N = 4, n = 2$, and $\alpha_1, \alpha_2, \alpha_3, \alpha_4$ be distinct. One gets Figure 3.6 for Δ. In all three regions of this figure the D–H polynomials are homogeneous quadratic polynomials. In the exterior regions these polynomials are monomials whose level sets are straight lines parallel to the exterior sides. If one goes into the interior region from the side parallel to α_2, the jump term that one has to add is a monomial whose level sets are straight lines parallel to α_2. If one goes into the interior region from the side parallel to α_3, the jump term that one has to add is a monomial whose level sets are straight lines parallel to α_3. (Exercise: Determine the coefficients of these jump terms using the fact that one has to get the same answer whether one goes into the interior region from the left or the right.)

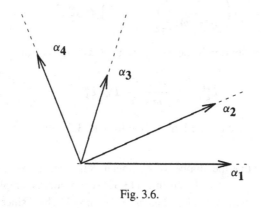

Fig. 3.6.

One important consequence of (3.33) is the following:

Theorem 3.2.11 f *is continuous near* W^S *if* $m > 1$, *and if* $m > k + 1$, f *is* k *times differentiable near* W^S.

3.3 Computing the D–H Polynomials: The "Heckman" Formula

Let's now come back to the situation we were considering in Section 3.1: G an n-torus, M a compact symplectic manifold, and

$$\kappa \colon G \to \mathrm{Symplecto}(M)$$

a Hamiltonian action of G on M. As in Section 3.1, we will denote by

$$\Phi: M \to \mathfrak{g}^*$$

the moment map associated with κ and by υ_Φ the push-forward with respect to Φ of the canonical symplectic measure on M. Let f be the Duistermaat–Heckman function, that is, the Radon–Nikodym derivative of υ_Φ with respect to Lebesgue measure on $\mathfrak{g}^*$. The goal of this section will be to derive a formula for f similar to the Kostant–Heckman formula which we discussed in the first paragraph of Section 3.1. Unfortunately, this formula will only make sense when the fixed-point set of G is a finite set, so we will henceforth assume that this is the case. Before stating this formula in its full generality, we will first describe what a "piece" of this formula looks like. As we pointed out in Section 3.1, the support Δ of υ_Φ is the convex hull of the image of the fixed-point set M_G of G; so, in particular, Δ is a polytope, and its vertices are images of points in M_G. However, there may be points in M_G that don't correspond to vertices of Δ: It is possible for the image of some fixed point to be contained in the interior of the convex hull of the remaining fixed points. In fact, a necessary and sufficient condition for a fixed point p to correspond to a vertex of Δ is the following: Let τ_p be the linear isotropy representation of G on the tangent space to M at p. Then there exists a G-invariant neighborhood U of p in M, a G-invariant neighborhood U_0 of the origin in $T M_p$, and a symplectomorphism

$$h: (U, p) \to (U_0, 0) \tag{3.37}$$

intertwining κ and τ_p. (For a proof of this "G-equivariant Darboux theorem" see [GS8].) Now let

$$\alpha_{p,i}, \quad i = 1, \dots, d, \tag{3.38}$$

be the weights of the representation of G on $T M_p$, and let $z_1, \dots, z_N$ be a system of complex coordinates in $T M_p$ compatible with the decomposition of $T M_p$ into weight spaces. The moment map Φ, restricted to U, is equal to $\Phi_0 \circ h + \Phi(p)$, where

$$\Phi_0(z_1, \dots, z_N) = \sum \alpha_{p,i} |z_i|^2. \tag{3.39}$$

(Compare with (3.19).) Thus the image of U in $\mathfrak{g}^*$ is the intersection of a neighborhood of $\Phi(p)$ with the cone

$$\{\Phi(p) + \sum s_i \alpha_{p,i} \mid s_1, \dots, s_N \geq 0\}, \tag{3.40}$$

and so $\Phi(p)$ will be a vertex of Δ, at least locally, iff (3.40) is a *proper* cone, that is, iff there exists $\xi \in \mathfrak{g}$ such that

$$\langle \alpha_{p,i}, \xi \rangle > 0 \tag{3.41}$$

for all i. One can, in fact, prove a good deal more. Using some global properties of the moment mapping, one can show that if $\Phi(U)$ is contained in the cone (3.40), then $\Phi(M)$ is contained in this cone, and for a sufficiently small neighborhood O of $\Phi(p)$, $\Phi^{-1}(O) \subseteq U$; so if (3.41) holds, $\Phi(p)$ really is a vertex of Δ, not just a vertex locally. (See [At] or [GS2].)

Let's now assume that a ξ satisfying (3.41) exists. Then, by Lemma 3.2.2, $\Phi \colon U \to \mathfrak{g}^*$ is proper; so the measure υ_Φ is identical in a neighborhood of $\Phi(p)$ to the measure that we studied in the previous section. In particular:

Theorem 3.3.1 *In a neighborhood of $\Phi(p)$, the Duistermaat–Heckman function is equal to*

$$ f_p(\mu + \Phi(p)) = \text{volume}\left\{ s_1, \ldots, s_N \geq 0 \ \Big| \ \sum s_i \alpha_{p,i} = \mu \right\}. \qquad (3.42) $$

Notice, by the way, that since f is a polynomial on each of the subregions Δ_i of Δ one can take the neighborhood on which $f(\mu)$ is equal to $f_p(\mu) + \Phi(p)$ to be a lot larger than $\Phi(U)$: One can take it to be the union of all the Δ_is whose closures contain $\Phi(p)$.

The moral of the foregoing discussion is that one can completely describe the measure υ_Φ in the vicinity of the vertices of Δ using nothing more than the relatively elementary results of the previous section and the G-equivariant Darboux theorem. Our goal, however, in this section is to determine υ_Φ not only in these exterior regions but inside all the Δ_is. For this we will need one of the deeper results of the Duistermaat–Heckman theory: the *exact stationary phase formula*. This formula, it turns out, can be viewed as a formula for the Fourier transform of υ_Φ on the complement of its singular support. From this we will by a kind of Tauberian argument be able to get υ_Φ itself. We will give a careful description of this formula later; but before we do so, let's first recall what the lemma of stationary phase in its usual form is about.

Let M be a compact n-dimensional manifold, μ a smooth, nonvanishing measure on M, and $\psi \colon M \to \mathbb{R}$ a smooth function having only a finite number of critical points, all of them nondegenerate. Then the lemma of stationary phase in its usual form is a recipe for evaluating the oscillatory integral

$$ \int e^{i\lambda\psi} \, d\mu $$

in terms of data at the critical points. More explicitly, it says that for λ large, this integral is equal to

$$ \lambda^{-n/2} \sum_p c_p e^{(i\pi/4)\,\text{sgn}(p)} \exp i\psi(p)\lambda \qquad (3.43) $$

modulo an error term of order $O(\lambda^{-n/2-1})$, where the sum is taken over the critical points. The c_ps and sgn(p)s in this sum are defined as follows. (See [GS1], chapter I.) The Hessian of ψ at p is a nondegenerate quadratic form on the tangent space to M at p, so one can choose a basis $v_1, \ldots, v_n$ for the tangent space such that with respect to this basis

$$d^2\psi(v_i, v_j) = \epsilon_i \delta_{ij},$$

where $\epsilon_i = \pm 1$. The number of $+1$'s minus the number of -1's is sgn(p) (i.e., the signature of $d^2\psi$), and c_p is the quantity $(2\pi)^{n/2}\mu(v_1, \ldots, v_n)$. In particular, let M be the symplectic manifold that we've been considering and μ the canonical symplectic measure, and let ψ be a component of the moment mapping; that is, let ξ be an element of g, and let $\psi = \phi^\xi = \langle \Phi, \xi \rangle$. It is easy to see that ϕ^ξ has nondegenerate critical points if and only if the conditions

$$\langle \alpha_{p,i}, \xi \rangle \neq 0 \tag{3.44}$$

are satisfied for all fixed points p of G, the $\alpha_{p,i}$s being as in (3.38), and, if these conditions are satisfied, that the critical points of ϕ^ξ coincide with these fixed points. Here is a proof of this fact. First we prove:

Lemma 3.3.2 *Let $p_1, \ldots, p_r$ be the fixed points of G and $U_1, \ldots, U_r$ neighborhoods of these fixed points. Let q be* any *point in M. Then there exists a point q_i in some U_i having the same stabilizer group as q.*

Proof. Let H be the stabilizer group of q. Let M_H be the set of points in M whose stabilizer group contains H, and let X be the connected component of M_H containing q. Then X is a compact symplectic submanifold of M (cf. [GS8], p. 203), and the torus G/H acts on it in a Hamiltonian fashion. Let X^0 be the set of points in X where this action is free. This set is an open set (in X), and since it contains q, it is nonempty; therefore, it is also *dense*. (Recall that if a compact group acts freely at any point then the set where its action is free is open and dense. This fact is an easy consequence of the principal orbit type theorem; see, for example, [Kaw].) The Hamiltonian action of a torus on a compact symplectic manifold always has fixed points (this is a trivial consequence of the convexity theorem); so there exists a point p' in X that is fixed by G/H, and arbitrarily close to this point there exists a point q' in X^0. However, p' has to be one of the p_is and we can choose q' to be a point q_i in the corresponding U_i. $\square$

Now suppose that ξ satisfies the conditions (3.41). Let q be a critical point of ϕ^ξ, and let H be its stabilizer group. This group contains the one-parameter subgroup of G generated by ξ. Let U_i be a G-invariant neighborhood of p_i on which the action of G is equivalent to the linear action of G on T_{p_i}. By the lemma there exists a point q_i in some U_i having stabilizer group H. Since H contains the group generated by ξ, q_i is a critical point of ϕ^ξ. However, the conditions (3.41) ensure that the *only* critical point of ϕ^ξ in U_i is p_i. Hence $H = G$ and q is a fixed point of G.

In fact, by (3.39) the Hessian of ϕ^ξ at p is

$$\sum_i \langle \alpha_{p,i}, \xi \rangle |z_i|^2 \qquad (3.45)$$

(and it's clear that this quadratic form is nondegenerate if and only if the conditions (3.44) are satisfied). From (3.39) one computes for the contribution of p to (3.43)

$$c_p = \left| \left(\prod_i \langle \alpha_{p,i}, \xi \rangle \right)^{-1} \right|$$

and

$$e^{\frac{i\pi}{4} \operatorname{sgn}(p)} = e^{\frac{i\pi}{2}\sigma} = i^\sigma,$$

where σ is the number of is such that $\alpha_{p,i}(\xi) > 0$, minus the number of is such that $\alpha_{p,i}(\xi) < 0$. Thus the stationary phase formula for ϕ^ξ reduces to

$$\int e^{i\lambda\phi^\xi} d\mu = \frac{i^\sigma}{\lambda^N} \sum \frac{e^{i\langle \Phi(p), \xi \rangle \lambda}}{\prod \langle \alpha_{p,i}, \xi \rangle} + O(\lambda^{-N-1}), \qquad (3.46)$$

for all ξ that satisfy (3.44). The exact stationary phase formula of Duistermaat–Heckman says that the error term on the right is *identically zero*; that is, setting $\lambda = 1$, the identity

$$\int e^{i\phi^\xi} d\mu = i^N \sum \frac{e^{i\langle \Phi(p), \xi \rangle}}{\prod \langle \alpha_{p,i}, \xi \rangle} \qquad (3.47)$$

holds on the nose. We will give a proof of this remarkable result in Appendix 3.B to this chapter. For the moment we will just say a few words about how one can "almost" prove (3.47) just using the G-equivariant form of the Darboux theorem. Namely, by the Darboux theorem, one can write the integrand on the left-hand side as

$$e^{i\lambda \sum \langle \alpha_{p,i}, \xi \rangle |z_i|^2} (1/2\pi i)^N dz \wedge d\bar{z}$$

in the neighborhood of the fixed point. Thus, by a partition-of-unity argument, one can write (3.46) as a sum of the integrals

$$\int e^{i\lambda \sum \langle \alpha_{p,i}, \xi \rangle |z_i|^2} (1/2\pi i)^N dz \wedge d\bar{z} \qquad (3.48)$$

(one such integral for each fixed point) and an integral of the form

$$\int \rho e^{i\lambda \phi^{\xi}} d\mu,$$

where ρ is identically zero in the neighborhood of the critical set. By elementary Fourier analysis, the second expression is of order $O(\lambda^{-\infty})$ in λ for λ large, and, by elementary calculations, one can show that the sum of the terms (3.48) is equal to the first term on the right-hand side in (3.46). Thus it takes no effort at all to improve the $O(\lambda^{-N-1})$ in (3.46) to an $O(\lambda^{-\infty})$. The hard part of the proof of (3.47) is getting rid of this innocuous-looking $O(\lambda^{-\infty})$.

We will now describe the "Heckman" formula alluded to at the beginning of this section. Consider the subset of $\mathfrak{g}$ consisting of the union of the hyperplanes

$$\langle \alpha_{p,i}, \xi \rangle = 0.$$

The complement of this set has several connected components, and we will fix, once and for all, one of these components and call it our *positive Weyl chamber*. (The formula we are in the course of deriving will actually be a collection of formulas, one for each choice of a positive Weyl chamber.)

Next we will define, for each fixed point p, a *renormalized* set of weights. Fix a vector ξ in the positive Weyl chamber, and let

$$\alpha_{p,i}^{w} = \epsilon_i \alpha_{p,i}, \tag{3.49}$$

where $\epsilon_i = +1$ or -1 depending on whether $\langle \alpha_{p,i}, \xi \rangle$ is positive or negative. In addition, let w_p be the number of ϵ_is that are negative. Associated with this set of renormalized weights, we will define a measure, on $\mathfrak{g}^*$, as follows. Let $\mathbb{R}_+^N$ be the positive orthant in $\mathbb{R}^N$, that is, the set of points $(s_1, \dots, s_N)$ with $s_1 \geq 0, \dots, s_N \geq 0$, and let

$$L_p \colon \mathbb{R}_+^N \to \mathfrak{g}^*$$

be the map

$$L_p(s_1, \dots, s_N) = (s_1 \alpha_{p,1}^{w} + \dots + s_N \alpha_{p,N}^{w}) + \Phi(p).$$

We will define υ_p to be the "push-forward"

$$\upsilon_p = (L_p)_* \, ds,$$

ds being Lebesgue measure on $\mathbb{R}_+^N$. Recall from the previous section that

$$\upsilon_p = f_p \, dx,$$

f_p being the "Heckman partition function"

$$f_p(\mu + \Phi(p)) = \text{volume} \left\{ s_1 \geq 0, \dots, s_N \geq 0 \,\middle|\, \sum s_i \alpha_{p,i}^{w} = \mu \right\}.$$

Theorem 3.3.3 *The Duistermaat–Heckman measure υ_Φ is equal to the alternating sum over the fixed-point set of G:*

$$\upsilon_\Phi = \sum_p (-1)^{w_p} \upsilon_p. \tag{3.50}$$

Before we prove this theorem, we will describe a simple example that helped to explain (for us) the mechanism that makes it work: Consider a region in the plane that, as in the following figure, has a quadrilateral boundary. A theorem of Delzant [Del] says that, modulo some mild rationality hypotheses on the angles at $\nu_1, \ldots, \nu_4$, there exists a compact four-dimensional symplectic manifold and a Hamiltonian action of T^2 on it with four fixed points. Moreover, one can arrange that they get mapped by the moment map onto the ν_is, and that the measure υ_Φ is just standard Lebesgue measure. (Indeed, Delzant's theorem says that there is a *unique* symplectic manifold, up to symplectomorphism, with these properties.) Let's now see what our theorem says about this example: To begin with, it says that Lebesgue measure on the foregoing region is equal to the sum (3.50) over the four fixed points. But if ξ is a vector pointing in the direction of the first quadrant, then the first measure in the preceding sum is Lebesgue measure supported on the shaded cone indicated in Figure 3.7. The second measure in the sum (3.50) is Lebesgue measure on the shaded cone in Figure 3.8. The third measure in the sum (3.50) is Lebesgue measure on the shaded region in Figure 3.9, and the fourth measure in the sum (3.50) is Lebesgue measure on the shaded region in Figure 3.10. The two sides of the cone in Figure 3.7 are oriented in the same direction as those of the parallelogram; so the first sign in the sum (3.50) is positive. We leave you to convince yourself that the second and third signs are negative and the fourth sign positive. With these assignments of sign, it's clear that the total sum is Lebesgue measure on the original region.

We turn now to the proof of Theorem 3.3.3. For the moment fix a point p in the fixed-point set, and set

$$\alpha_i = \alpha_{p,i}^w, \quad i = 1, \ldots, N,$$

and

$$\nu = \Phi(p).$$

Consider the function

$$g(\xi) = i^N e^{i\langle \nu, \xi \rangle} \prod_{i=1}^N \langle \alpha_i, \xi \rangle^{-1}. \tag{3.51}$$

Let O be the open subset of $\mathfrak{g}$ obtained by deleting the hyperplanes $\langle \alpha_i, \xi \rangle = 0$.

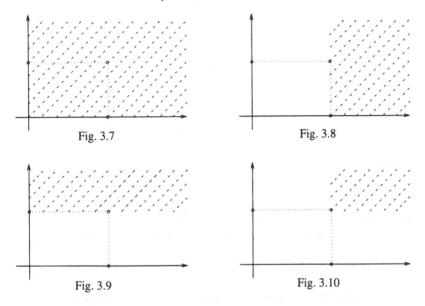

Fig. 3.7   Fig. 3.8

Fig. 3.9   Fig. 3.10

Then g is a well-defined rational function on O, and there are many ways to extend it to a generalized function on all of $\mathfrak{g}$. We will prove, however:

Proposition 3.3.4 *Let ξ be a vector in the positive Weyl chamber. Then there is one and just one tempered distribution h on $\mathfrak{g}$ with the properties that h is equal to g on O and that the Fourier transform of h is properly supported in the half-space $\{\alpha \in \mathfrak{g}^* : \langle \alpha, \xi \rangle \geq c\}$ for some constant c.*

Proof. Without loss of generality we can assume that $\nu = 0$ (in which case we can take the constant c to be zero). The existence of one such distribution is easy. Let υ be the measure defined by (3.22). We showed that this measure satisfies the differential equation

$$D_{\alpha_1} \cdots D_{\alpha_N} \upsilon = \delta_0,$$

so the Fourier transform of this measure satisfies

$$\prod \langle \alpha_i, \xi \rangle \hat{\upsilon}(\xi) = 1,$$

which reduces to (3.51) on O.

As for the uniqueness, suppose there are two tempered distributions, h_1 and h_2, with the aforementioned properties. Let h_3 be their difference. Then h_3 is supported on the union of the hyperplanes $\langle \alpha_i, \xi \rangle = 0$, and, therefore, h_3

multiplied by some large power of $\prod \langle \alpha_i, \xi \rangle$ has to vanish. Hence its Fourier transform has to satisfy the differential equation,

$$(D_{\alpha_1} \cdots D_{\alpha_N})^r f = 0,$$

for r large, and is supported on the half-space $\langle \mu, \xi \rangle \geq 0$. From this it follows by Lemma 3.2.5 and induction on N that $f \equiv 0$. Hence the Fourier transform of h_3 is identically zero, and consequently so is h_3 itself. $\qquad \square$

To prove Theorem 3.3.3 notice that by construction the Fourier transform of the left-hand side of (3.50) is equal to the Fourier transform of the right on O. However, by the Fourier inversion formula, the Fourier transforms of these Fourier transforms are both supported in a fixed half-space. Hence the same argument as that we have just given shows that they have to be identical.

3.4 Is There a "Kostant" Formula Corresponding to the "Heckman" Formula?

We will show in this section that the answer to this question is yes. We will have to assume, however, that the symplectic action of G on M, that is,

$$\kappa \colon G \to \text{Symplecto}(M),$$

can be "quantized." As we pointed out in Section 3.1, there is no firm consensus about what this term means; however, for the duration of Section 3.4, we will take it to mean that the following are true.

(a) κ can be *prequantized*: Suppose that the cohomology class defined by the symplectic form on M belongs to the image of the canonical mapping $i \colon H^2(M, \mathbb{Z}) \to H^2(M, \mathbb{R})$, that is, that it is an *integral* cohomology class. Then, by [K2], there exists a Hermitian line bundle L sitting over M and a connection ∇ on L whose curvature form is ω_M. Moreover, if M is simply connected, L and ∇ are unique up to isomorphism. Furthermore (loc. cit.), one gets a canonical representation of $\mathfrak{g}$ on the space of sections of L. For each $\xi \in \mathfrak{g}$, let $\xi^\#$ be the vector field on M associated with ξ, and let

$$D_\xi = \nabla_{\xi^\#} + i \langle \Phi, \xi \rangle. \qquad (3.52)$$

Then the map $\xi \to D_\xi$ is a Lie algebra homomorphism. Following [K2] we will say that κ can be prequantized if there is a representation of G on sections of L that is given infinitesmally by (3.52).

(b) M possesses a positive-definite G-invariant polarization, that is, there exists a G-invariant Kähler structure on M compatible with its symplectic structure. In this case L aquires naturally the structure of a holomorphic line bundle, and the holomorphic structure on L is compatible with the action of G defined by (3.52).

Given (a) and (b) we will, again following [K2], define the quantization of κ to be the natural representation ρ of G on the space of holomorphic sections of L. Our starting point for the "Kostant" formula to follow will be a formula for the character of ρ that is very similar to the exact stationary phase formula described in the previous section. As in the previous section, we will assume that the fixed-point set M_G is finite and that for each $p \in M_G$, the weights of the isotropy representation of G on the tangent space to M at p are given by (3.38). Let ξ be an element of $\mathfrak{g}$ satisfying (3.44), and consider the trace of $\rho(\exp \xi)$. By the Atiyah–Bott fixed-point formula [AB], this trace is equal to the sum over M_G:

$$\sum_{p \in M_G} \frac{\mathrm{tr}\{\kappa_p(\exp \xi) : L_p \to L_p\}}{\prod(1 - e^{i\langle \alpha_{p,k}, \xi \rangle})},$$

κ_p being the isotropy representation of G on the fiber L_p. By (3.52) the trace of $\kappa_p(\exp \xi)$ is equal to $e^{i\langle \Phi(p), \xi \rangle}$; so we get the formula

$$\mathrm{tr}\,\rho(\exp \xi) = \sum_{p \in M_G} \frac{e^{i\langle \Phi(p), \xi \rangle}}{\prod(1 - e^{i\langle \alpha_{p,k}, \xi \rangle})}. \tag{3.53}$$

Before exploiting this formula, we will pause for a moment to point out some similarities between it and (3.47). In both formulas, the sum is over the fixed-point set of G and the numerator of the pth summand is the same. Moreover, for ξ small, the denominator of the pth summand in (3.53) is approximately equal to

$$(-i)^N \prod \langle \alpha_{p,k}, \xi \rangle,$$

which is identical to the denominator of the pth summand in (3.47). Thus we've proved:

Proposition 3.4.1 *For ξ small, the trace of $\rho(\exp \xi)$ is approximately equal to $\hat{\upsilon}_\Phi(\xi)$.*

This result explains why the asymptotic properties of the multiplicity diagrams can be described by a "central limit theorem" involving the measure υ_Φ. (See Example 3.10 of Section 3.1.) Let's now consider the Fourier transforms of the right- and left-hand sides of (3.53). Let α be a lattice point in $\mathfrak{g}^*$ and let $*(\alpha, \rho)$ be the multiplicity with which the weight corresponding to α occurs in ρ. Then the left-hand side of (3.53) is the sum

$$\sum *(\mu, \rho)e^{i\langle\mu,\xi\rangle}$$

and its Fourier transform is a finite sum of delta functions:

$$\sum *(\mu, \rho)\delta(\lambda - \mu). \tag{3.54}$$

On the other hand to compute the Fourier transform of the right-hand side of (3.53), we have to know how to compute the Fourier transforms of each of the individual terms on the right-hand side. The situation here is similar to the situation we encountered in the previous section in trying to compute the right-hand side of (3.47). There is some ambiguity about the definition of this Fourier transform since the pth summand in (3.53) is defined only if

$$\langle\alpha_{p,k}, \xi\rangle \notin \mathbb{Z}$$

for all k and p. As in the previous section, we will deal with this ambiguity by *renormalizing* the $\alpha_{p,k}$s, that is, we will fix once and for all a "positive Weyl chamber" in the set

$$\langle\alpha_{p,k}, \xi\rangle \neq 0 \text{ for all } k \text{ and } p$$

and define our renormalized $\alpha_{p,k}$s to be

$$\alpha_{p,k}^w = \epsilon_k \alpha_{p,k}, \tag{3.55}$$

where $\epsilon_k = +1$ or -1 depending on whether $\alpha_{p,k}$ is greater or less than zero for all ξ in the positive Weyl chamber. After some juggling one can write the pth term in (3.53) as

$$(-1)^{w_p} \frac{e^{i\langle\Phi(p)+\delta_p^w-\delta_p, \xi\rangle}}{\prod(1 - e^{i\langle\alpha_{p,k}^w, \xi\rangle})}, \tag{3.56}$$

where $\delta_p = 1/2\sum\alpha_{p,k}$ and $\delta_p^w = 1/2\sum\alpha_{p,k}^w$. Now let's replace each factor in the denominator of this expression by the corresponding geometric series

$$\sum_{m=0}^{\infty} e^{i m\langle\alpha_{p,k}^w,\xi\rangle}$$

and get for (3.56) the formula

$$(-1)^{w_p} \sum N_p(\mu)e^{i\langle\mu+\Phi(p)+\delta_p^w-\delta_p,\xi\rangle}, \tag{3.57}$$

where $N_p(\mu)$ is the number of N-tuples of nonnegative integers $(k_1, \ldots, k_N)$ satisfying the equation

$$\mu = \sum k_i \alpha_{p,i}^w.$$

We will call N_p the *partition function* associated with the point p. Consider now the Fourier transform of (3.57). This is the sum

$$(-1)^{w_p} \sum N_p(\mu)\delta(\lambda - \mu - \Phi(p) - \delta_p^w + \delta_p). \tag{3.58}$$

Notice that both (3.58) and (3.54) have support in a fixed half-space

$$H_\xi = \{\lambda \in \mathfrak{g}^* \mid \langle \lambda, \xi \rangle \geq c\},$$

with ξ lying in the positive Weyl chamber and c being a fixed constant. (For (3.58) this is a consequence of the fact that N_p is supported in the cone

$$\left\{ \sum s_i \alpha_{p,i}^w \;\middle|\; s_1 \geq 0, \ldots, s_N \geq 0 \right\},$$

and for (3.54) it is merely a consequence of the fact that (3.54) is compactly supported.) Therefore, as in Section 3.3, this implies that the expression (3.54) is equal *on the nose* to the sum over $p \in M_G$ of the expressions (3.58) (both being viewed as distributional functions of λ). Let us now compare both sides of this equation. Setting $\nu = \mu + \Phi(p) + \delta_p^w - \delta_p$, we can rewrite (3.58) as

$$(-1)^{w_p} \sum N_p(\nu - \Phi(p) - \delta_p^w + \delta_p)\delta(\lambda - \nu).$$

Thus, summing over p and comparing the coefficients of $\delta(\lambda - \nu)$, we get the "Kostant" formula:

$$* (\nu, \rho) = \sum_{p \in M_G} (-1)^{w_p} N_p(\nu - \delta_p^w - (\Phi(p) - \delta_p)). \tag{3.59}$$

3.4.1   The Kostant Multiplicity Formula for the Spin Representation of SO(2n)

In this subsection we present our version of the Kostant multiplicity formula for the spin representation of $SO(2n)$. This version appears to be quite different from the usual version of the Kostant multiplicity formula.

Let X be the (co)adjoint orbit of $SO(2n)$ consisting of all skew-adjoint $2n \times 2n$ matrices J with $J^2 = -\text{Id}$ and $\det J = 1$. As a homogeneous space

$$X = SO(2n)/U(n).$$

(Several equivalent descriptions of X are given in the Appendix to this section.) Let T be the Cartan subgroup of $SO(2n)$ (or, alternatively, the Cartan subgroup of $U(n)$; both these groups have the same Cartan subgroup). Let p_0 be the identity coset in X. The representation of T on the tangent space $T_{p_0}X$ is the restriction to T of the representation of $U(n)$ on $\Lambda^2\mathbb{C}^n$. Indeed, the tangent space to X at a point J can be identified with the set of all skew-adjoint matrices satisfying $AJ = -JA$, while J defines a complex structure on $\mathbb{R}^{2n}$. It is easy to check that the $\mathbb{C}$-valued bilinear form $\langle Ax, y \rangle + i\langle JAx, y \rangle$ is complex bilinear with respect to J and antisymmetric. For the identity coset this gives an identification of the tangent space with $\Lambda^2\mathbb{C}^n$, for the complex structure associated with $U(n)$. The weights of this representation are

$$\theta_i + \theta_j, \quad i < j, \tag{3.60}$$

where $\theta_1, \ldots, \theta_n$ are the weights of the representation of T on $\mathbb{C}^n$. (Notice that these weights are a basis for the weight lattice $\mathbb{Z}_T$ of T.)

Let Σ be the abelian component of the Weyl group of $SO(2n)$. The group Σ acts in a simply transitive fashion on the fixed-point set X_T, so we get an identification:

$$\Sigma \to X_T, \quad a \mapsto ap_0.$$

The action of Σ on $\mathbb{Z}_T$ is faithful and takes the form

$$a\theta_i = \pm\theta_i, \quad i = 1, \ldots, n. \tag{3.61}$$

In fact the sign convention in (3.61) gives rise to an isomorphism between Σ and the group

$$\{(\epsilon_1, \ldots, \epsilon_n) \mid \epsilon_i = \pm 1, \ \epsilon_1 \cdots \epsilon_n = 1\}. \tag{3.62}$$

Therefore, by (3.61), the weights of the isotropy representation of T on $T_{ap_0}X$ are

$$\theta_i^a + \theta_j^a = \epsilon_i\theta_i + \epsilon_j\theta_j, \quad i < j. \tag{3.63}$$

If we set

$$S = \{j \mid \epsilon_j = 1\} \tag{3.64}$$

these can be tabulated in the form:

$$\begin{array}{lll}
\theta_i + \theta_j, & i < j & \text{and} \quad i, j \in S, \\
-\theta_i - \theta_j, & i < j & \text{and} \quad i, j \in S^c, \\
-\theta_i + \theta_j, & i < j & \text{and} \quad i \in S^c, \ j \in S, \\
\theta_i - \theta_j, & i < j & \text{and} \quad i \in S, \ j \in S^c.
\end{array} \tag{3.65}$$

Let us choose as a polarizing vector the vector ξ defined by

$$\theta_i(\xi) = i, \quad i = 1, \ldots, n. \tag{3.66}$$

Then the polarized weights at $p = ap_0$ are

$$
\begin{array}{llll}
\theta_i + \theta_j, & i < j & \text{and} & i, j \in S, \\
\theta_i + \theta_j, & i < j & \text{and} & i, j \in S^c, \\
-\theta_i + \theta_j, & i < j & \text{and} & i \in S^c, \ j \in S, \\
-\theta_i + \theta_j, & i < j & \text{and} & i \in S, \ j \in S^c;
\end{array} \tag{3.67}
$$

that is, the first and third lines in (3.67) are the same as in (3.65) and the second and fourth lines are multiplied by -1. Let's first of all count sign changes. By the previous remark the total number of sign changes is

$$
\#\{(i, j), \ i < j \mid i, j \in S^c\}
$$

plus

$$
\#\{(i, j), \ i < j \mid i \in S, \ j \in S^c\},
$$

or in other words

$$
\#\{(i, j) \mid i < j, \ j \in S^c\},
$$

which is just

$$
\sum_{i \in S^c} (j - 1).
$$

Since $j - 1$ is even if j is odd and vice versa, the number of sign changes is

$$
(-1)^p = (-1)^{\kappa(S^c)}, \tag{3.68}
$$

where, for any subset A of $\{1, \ldots, n\}$, $\kappa(A)$ is the number of *even* elements in A.

Next let us compute $\delta_p - \delta_p^w$. This is the sum of all the terms on the second row of (3.65) plus the terms on the fourth row, that is,

$$
- \sum_{i<j, i, j \in S^c} \theta_i + \theta_j + \sum_{i<j, i \in S, j \in S^c} \theta_i - \theta_j,
$$

which can be rewritten as

$$
\sum_{j \in S^c} \left(\sum_{i<j} -\theta_j + \epsilon_i \theta_i \right)
$$

or alternatively as

$$
\frac{1}{2} \sum_j (1 - \epsilon_j) \sum_{i<j} -\theta_j + \epsilon_i \theta_i. \tag{3.69}
$$

This we will break up into the sum of two terms

$$\frac{1}{2} \sum_j \left(\sum_{i<j} \epsilon_i \theta_i + \epsilon_j \theta_j \right) \tag{3.70}$$

and

$$-\frac{1}{2} \sum_j \left(\sum_{i<j} \theta_j + \epsilon_i \epsilon_j \theta_i \right). \tag{3.71}$$

The first term is symmetric in (i, j) so it can be written in the form

$$\frac{1}{4} \sum_{i,j} (\epsilon_i \theta_i + \epsilon_j \theta_j) - \sum 2\epsilon_i \theta_i,$$

which is just

$$\frac{1}{4}(n-1) \sum \epsilon_i \theta_i. \tag{3.72}$$

The second term can be written as

$$-\frac{1}{2} \sum (j-1)\theta_j - \frac{1}{2} \sum_j \left(\sum_{i<j} \epsilon_i \epsilon_j \theta_i \right).$$

Reversing the summation in i and j the last expression becomes

$$-\frac{1}{2} \sum_i \epsilon_i \left(\sum_{j>i} \epsilon_j \theta_j \right),$$

so if we set

$$\tau_i = \sum_{j>i} \epsilon_j \tag{3.73}$$

we get for the second term the formula

$$\frac{1}{2} \sum (\epsilon_i \tau_i)\theta_i - \delta,$$

where

$$\delta = \frac{1}{2} \sum (i-1)\theta_i. \tag{3.74}$$

Finally, adding this to (3.72) we get the formula

$$\delta_p - \delta_p^w = \left(\sum \frac{\epsilon_i}{2} \left(\frac{n-1}{2} - \tau_i \right) \theta_i \right) - \delta. \tag{3.75}$$

It is easy to check that

$$\Phi(p) = \sum \frac{\gamma \epsilon_i}{2} \theta_i$$

at $p = ap_0$, γ being a fixed integer which depends on the choice of the symplectic form on X. Hence we get for the "shift" in the multiplicity formula

$$\delta_p - \delta_p^w + \Phi(p) = \left(\sum \frac{\epsilon_i}{2} \left(\frac{n-1}{2} + \gamma - \tau_i \right) \theta_i \right) - \delta. \qquad (3.76)$$

Thus we've computed the "sign changes" (3.68) and the "shifts" (3.76) and we
come finally to the main object of interest – the counting function N_p.

For each $\alpha = \sum a_i \theta_i$, $N_p(\alpha)$ is the number of ways of writing α as a sum of
polarized weights:

$$\sum_{i<j}{}' k_{ij}(\theta_i + \theta_j) + \sum_{i<j}{}'' k_{ij}(-\theta_i + \theta_j),$$

where the first sum is over all i, j for which $\epsilon_i \epsilon_j = +1$ and the second sum is
over all i, j for which $\epsilon_i \epsilon_j = -1$. (The k_{ij}s, of course, have to be nonnegative
integers.) If we make the convention that $k_{ij} = k_{ji}$ for $i \neq j$ and k_{ii} this can
be written as a single sum

$$\sum_{i<j} k_{ij}(\epsilon_i \epsilon_j \theta_i + \theta_j),$$

which is the same as

$$\sum_i \left(\sum_{j>i} k_{ij} \epsilon_i \epsilon_j \right) \theta_i + \sum_i \left(\sum_{j<i} k_{ij} \right) \theta_i.$$

Thus setting

$$n_{ij} = \begin{cases} \epsilon_i \epsilon_j k_{ij}, & i < j, \\ k_{ij}, & i \geq j, \end{cases}$$

we have proved:

Theorem 3.4.2 *If* $\alpha = \sum a_i \theta_i$ *then* $N_p(\alpha)$ *is the set of* $n_{ij} \in \mathbb{Z}$, $1 \leq i, j \leq n$,
satisfying

$$a_i = \sum_{j=1}^{n} n_{ij} \qquad (3.77)$$

and satisfying also the auxiliary conditions

$$\begin{aligned} n_{ij} &\geq 0 \, for \, i > j, \\ n_{ii} &= 0, \\ n_{ij} &= \epsilon_i \epsilon_j n_{ji}. \end{aligned} \qquad (3.78)$$

Notice that the numbers (3.77) are just the row sums of the matrix (n_{ij}) so $N_p(\alpha)$ is just the number of matrices $N = (n_{ij})$, $1 \le i, j \le n$, with prescribed row sums (3.77) satisfying the conditions (3.78). Notice also that the third of the conditions (3.78) just says that

$$N^t = C_\epsilon N C_\epsilon, \qquad (3.79)$$

where C_ϵ is the diagonal matrix with entries $\epsilon_1, \ldots, \epsilon_n$ along the diagonal.

Concluding Remarks

1. N_{p_0} is just the "graph counting" function. Let $N = (n_{ij})$ be the adjacency matrix of a graph on n vertices. Then the degrees of the vertices are just the row sums of N, so N_{p_0} counts the number of graphs for which the degrees of the vertices are prescribed, that is, are just the a_is in the sums (3.77). (Note that for p_0, $\epsilon_1 = \cdots = \epsilon_n = 1$ so $n_{ij} = n_{ji}$.)

2. Here is a minor cosmetic change which makes the formula for the multiplicity function $m(\alpha)$ a little simpler. Given $\epsilon = (\epsilon_1, \ldots, \epsilon_n)$, let N^ϵ be the matrix

$$n_{ij} = \begin{cases} 1, & i > j, \\ 0, & i = j, \\ \epsilon_i \epsilon_j & i < j. \end{cases} \qquad (3.80)$$

Then the row sums of N^ϵ are

$$s_i = \sum_{j=1}^n n_{ij} = \sum_{i>j} 1 + \sum_{i<j} \epsilon_i \epsilon_j = i - 1 + \epsilon_i \tau_i.$$

Thus

$$\sum s_i \theta_i = \sum (i-1)\theta_i + \sum \epsilon_i \tau_i \theta_i.$$

Thus the row sums of $\frac{1}{2} N^\epsilon$ are $s_i/2$, where

$$\sum s_i/2\,\theta_i = \delta + \frac{1}{2} \sum \epsilon_i \tau_i \theta_i.$$

In other words, in view of the formulas (3.74)–(3.76), if N satisfies (3.78) and its row sums are $b_1, \ldots, b_n$, where

$$\sum b_i \theta_i = \delta_p - \delta_p^w + \Phi(p) - \alpha,$$

then the row sums of $N + \frac{1}{2} N^\epsilon$, call them $c_1, \ldots, c_n$, satisfy

$$\sum c_i \theta_i = \frac{1}{2} \left(\frac{n-1}{2} + \gamma \right) \left(\sum \epsilon_i \theta_i \right) - \alpha.$$

Hence to summarize:

Theorem 3.4.3 *If $p = a p_0$ and N^ϵ is the matrix (3.80) then the multiplicity*

$$N_p(\delta_p - \delta_p^w + \Phi(p) - \alpha)$$

is equal to the number of matrices N satisfying (3.78) such that the row sums $c_1, \ldots, c_n$ of $N + \frac{1}{2}N^\epsilon$ satisfy

$$\sum c_i \theta_i = \frac{1}{2}\left(\frac{n-1}{2} + \gamma\right)\left(\sum \epsilon_i \theta_i\right) - \alpha.$$

Appendix: Descriptions of $SO(2n)/U(n)$

In this appendix we give several (equivalent) descriptions of the coadjoint orbit $X = SO(2n)/U(n)$ and show that up to a scalar multiple it possesses just one $SO(2n)$ invariant symplectic structure.

As usual we will identify $\mathfrak{o}(2n)$ and $\mathfrak{o}(2n)^*$. The coadjoint orbit associated with the spin representation consists of all elements of $\mathfrak{o}(2n)$ that are conjugate to

$$\begin{pmatrix} J_2 & 0 & 0 & \cdots & 0 \\ 0 & J_2 & 0 & \cdots & 0 \\ 0 & 0 & J_2 & \cdots & 0 \\ \vdots & \vdots & \vdots & \ddots & \vdots \\ 0 & 0 & 0 & \cdots & J_2 \end{pmatrix}, \qquad (3.81)$$

where J_2 is the 2×2 matrix

$$\begin{pmatrix} 0 & -1 \\ +1 & 0 \end{pmatrix}.$$

Let us denote this orbit by X. Here are some other descriptions of it.

1. X is the set of all skew-adjoint $2n \times 2n$ matrices J for which $J^2 = -\mathrm{Id}$.
2. X is the set of all complex structures on $\mathbb{R}^{2n}$ that are compatible with the standard inner product, that is, satisfy

$$\langle Jx, Jy \rangle = \langle x, y \rangle. \qquad (3.82)$$

Proof that $1 \Leftrightarrow 2$. The identity (3.82) holds if and only if

$$\langle Jx, y \rangle + \langle x, Jy \rangle = 0, \qquad (3.83)$$

that is, if and only if J is skew-adjoint. $\qquad \square$

3. X is the Grassmannian of all isotropic n-dimensional subspaces of $\mathbb{C}^{2n}$ (with respect to the quadratic form $z_1^2 + \cdots + z_{2n}^2$).

Proof. Let W be an isotropic n-dimensional subspace of $\mathbb{C}^{2n}$; the $\mathbb{R}$-linear maps

$$A\colon W \to \mathbb{R}^{2n}, \quad w \mapsto Re\, w$$

and

$$B\colon W \to \mathbb{R}^{2n}, \quad w \mapsto Im\, w$$

are bijective. Set $J = BA^{-1}$. Let u be in $\mathbb{R}^n$ and let $v = Ju$. Then by definition $w =: u + iv$ is in W, so $-iw = v - iw$ is in W. Hence $Jv = -u$; that is, $J^2 u = -u$. Thus $J^2 = -\,\mathrm{Id}$.

Since W is isotropic,

$$\langle w, w \rangle = \langle u, u \rangle - \langle v, v \rangle + i(\langle u, v \rangle + \langle v, u \rangle) = 0.$$

Hence

$$\langle Ju, Ju \rangle = \langle u, u \rangle$$

and

$$\langle u, Ju \rangle = 0.$$

Thus J defines a complex structure on $\mathbb{R}^{2n}$ compatible with $\langle\,,\,\rangle$. Conversely if J defines such a complex structure then by reversing the previous argument one sees that the space

$$W_J = \{ u + iJu \mid u \in \mathbb{R}^{2n} \}$$

is an isotropic complex n-dimensional subspace of $\mathbb{C}^{2n}$. □

4. Finally, $O(2n)$ acts transitively on X and as a homogeneous space

$$X = O(2n)/U(n). \qquad (3.84)$$

Proof. $U(n)$ is the commutator of J in $O(2n)$; so this follows from the definition. □

Remark. It is clear from (3.82) that X consists of two disjoint pieces each of which is a coadjoint orbit of $SO(2n)$.

Since X is a coadjoint orbit it admits a symplectic structure. We will show that in fact up to scaling the symplectic structure is unique. To see this we first observe:

Lemma 3.4.4 *Let e be the identity coset of $O(2n)/U(n)$. Then the isotropy representation of $U(n)$ on the tangent space at e is the standard representation of $U(n)$ on $\Lambda^2\mathbb{C}^n$.*

Proof. This follows from our third description of X on the foregoing list.   □

Since the representation is $\mathbb{R}$-irreducible we obtain:

Corollary 3.4.5 *Up to scaling there is a unique $O(2n)$-invariant symplectic structure on X.*

3.5   Computing the D–H Polynomials: Formulas for the Jumps

In the exact stationary phase formula that we described in Section 3.3 we had to assume that the fixed-point set of G, that is, M_G, was a finite set. Duistermaat and Heckman derived in [DH2] a more general version of exact stationary phase that makes no assumptions at all about the fixed-point set of G. We will give a description of this result in the appendix. Unfortunately it is not very useful for the kinds of computations we are trying to do in this chapter, since it involves knowing quite a bit about the topology of the normal bundle of M_G. However, the "leading term" in their formula can be computed by a standard stationary-phase argument and is very useful for computing the "jumps" in the D–H polynomials across the walls of the Δ_is. It turns out that these jumps are given by essentially the same formulas as in the linear case. Here are the details.

Let Δ_+ and Δ_- be two adjacent Δ_is, let f_+ and f_- be the D–H polynomials associated with them, and let W be the $(n-1)$-dimensional wall separating them, oriented so that the normal vector ξ to W is pointing out of Δ_- and into Δ_+. We will show that the jump $f_+ - f_-$ at W is given by the formula

$$f_+ - f_- = \left(\prod_{i=1}^{m}\langle\alpha_i,\xi\rangle^{-1}\right) f_W \frac{(L_\xi - c)^{m-1}}{(m-1)!} \tag{3.85}$$

plus an error term of order $(L_\xi - c)^m$, the quantities α_i, f_W, L_ξ, and so forth being defined more or less as in the linear case. (Compare with (3.33).)

The precise definitions are as follows. As we already mentioned, ξ is the normal vector to W pointing into Δ_+. (Since W sits in $\mathfrak{g}^*$, we can think of ξ as an element of $\mathfrak{g}$.) By L_ξ we will mean the linear functional on $\mathfrak{g}^*$ associated with ξ, that is, the linear functional $\mu \to \langle\mu,\xi\rangle$.

The one-parameter subgroup of G generated by ξ is a circular subgroup, and we will denote it by S^1. To explain the other undefined quantities on the right-hand side of (3.85), we must first say a few words about the action of S^1 on M. To begin with, its fixed point set M_{S^1} is a symplectic submanifold of M, and Φ maps one (or more) of the connected components of M_{S^1} onto W. (For simplicity, we will assume that just one component is involved and denote this component by X.) The quotient group G/S^1 acts in a Hamiltonian fashion on X, and the moment map of this action is the restriction of Φ to X. The push-forward of the canonical symplectic measure on X by the moment map is a measure that lives on W, and the f_W in (3.85) is the Radon–Nikodym derivative of this measure with respect to the Lebesgue measure on W. (A small complication is "how to define Lebesgue measure on W." In principle it is only defined up to the choice of a positive constant; however, as we pointed out in Section 3.2, the choice of ξ fixes this constant in a canonical way. See (3.31) and the discussion following it.)

We still have to explain the "m" and the "$\langle \alpha_i, \xi \rangle$s" in (3.85). The integer m is, as in (3.33), the codimension of X divided by 2, and the $\langle \alpha_i, \xi \rangle$s are the weights of the isotropy representation of S^1 on the normal bundle of X. (Notice that, because X is connected, this representation has the same weights at every point.)

Proof of (3.85). Let H be an $(n-1)$-dimensional toral subgroup of G with the property that H and S^1 intersect in $\{e\}$, and let Φ_H be the moment mapping associated with the action of H on X. Φ_H is the composition of Φ with the canonical projection

$$\pi : \mathfrak{g}^* \to \mathfrak{h}^*,$$

and the restriction of π to W maps W bijectively onto $\mathfrak{h}^*$. Let α be a point on W whose image $\pi(\alpha)$ is a regular value of Φ_H. Associated to α are two symplectic manifolds: One can reduce M (viewed as an H-space) at $\pi(\alpha)$, and one can reduce X (viewed as a G/S^1-space) at α. Let us denote these reduced spaces by M_α and X_α. Notice that since the action of S^1 commutes with the action of H, there is an induced action of S^1 on M_α. Let Y_α be the preimage of $\pi(\alpha)$ with respect to Φ_H. Since $\pi(\alpha)$ is a regular value of Φ_H the action of H on Y_α is locally free; and, for simplicity, we will assume it is free. In this case Y_α is a principal H-bundle over M_α:

$$Y_\alpha \longrightarrow M_\alpha. \tag{3.86}$$

Let's restrict Φ to Y_α. By definition Φ maps Y_α onto the line $\alpha + \mathfrak{h}^0$ in $\mathfrak{g}^*$. Moreover, it is constant on the fibers of (3.86); so it induces on the base a mapping

$$\phi : M_\alpha \longrightarrow \alpha + \mathfrak{h}^0.$$

Composing this mapping with the mapping

$$\alpha + \mathfrak{h}^0 \longrightarrow \mathbb{R}, \qquad \alpha + \nu \longrightarrow \langle \alpha + \nu, \xi \rangle$$

we get a mapping of M_α into $\mathbb{R}$ which we will continue to denote by ϕ. It is easy to see that this is just the moment map associated with the S^1-action on M_α that we described earlier.

Lemma 3.5.1 *X_α is the set of critical points of the mapping ϕ lying on the critical level $\phi = c$.*

We will leave the proof of this as an easy exercise. Notice that, by definition, the value of f_W at α is the symplectic volume of X_α. So, by the lemma, we are reduced to proving (3.85) in the special case $G = S^1$; that is, we are reduced to proving the following (the manifold M in the theorem that follows being the "M_α" from before):

Theorem 3.5.2 *Let M be a compact symplectic manifold. Suppose S^1 acts on M in a Hamiltonian fashion with moment map $\phi \colon M \to \mathbb{R}$. Let $f(t)dt$ be the push-forward by ϕ of the symplectic measure on M, and let C_α be the critical set*

$$\{ m \in M \mid \phi(m) = \alpha, d\phi_m = 0 \}.$$

Then the jump in $f(t)$ at α is given by the formula:

$$f_+ - f_- = \text{volume}(C_\alpha) \left(\prod_{i=1}^k \alpha_i^{-1} \right) \frac{(t - \alpha)^{k-1}}{(k - 1)!} \tag{3.87}$$

plus an error term of order $O((t - \alpha)^k)$, the α_is being the weights of the representation of S^1 on the normal bundle of C_α.

Proof. Let μ be the symplectic volume form on M. It is clear that the Fourier transform of f is the integral

$$\int e^{is\phi} d\mu. \tag{3.88}$$

The contribution of C_α to the stationary-phase expansion of (3.88) is

$$\text{volume}(C_\alpha) \left(\prod_{j=1}^m \alpha_j^{-1} \right) s^{-m} e^{is\alpha} + O(s^{-m-1}). \tag{3.89}$$

(Here we are just using the standard stationary phase formula for "clean" phase functions. See, for instance, [Hör], page 222.)

Now let's inspect the function $f(t)$ a little bit more carefully. By the Duistermaat–Heckman theorem, $f(t)$ is a piecewise-polynomial function of the form

$$\sum x_I(t) P_I(t),$$

the $P_I(t)$s being polynomials in t, the Is being subintervals of the real axis whose endpoints are critical values of ϕ, and $x_I(t)$ being the characteristic function of the interval I. The Fourier transform of this sum is

$$\sum P_I \left(\frac{1}{i} \frac{d}{ds} \right) \frac{e^{i\beta_I s} - e^{i\alpha_I s}}{s},$$

β_I and α_I being the endpoints of the interval I. The critical value α lies on either one or two of these intervals depending on whether it is an interior critical value or extremal critical value. In either case, by comparing (3.89) with the singular terms involving α in the foregoing sum one easily deduces (3.87).   □

The formula (3.85) is particularly useful if the action of G/S^1 on X is a "Delzant" action, that is, if the dimension of X is as small as possible, namely, twice the dimension of the group G/S^1. In this case f_W is equal to a constant (which we will denote by c_W) and (3.85) becomes an exact formula:

$$f_+ - f_- = c_W \left(\prod_{i=1}^{N-n} \langle \alpha_i, \xi \rangle^{-1} \right) \frac{(L_\xi - c)^{N-n}}{(N-n)!}. \tag{3.90}$$

This formula turns out to be very useful for computing the D–H polynomials for Lie groups of low rank. We will give an account of the Duistermaat–Heckman theory for these groups in Chapter 5. However, to illustrate how useful (3.90) can be, we will give here a brief sketch of how to compute the D–H polynomials for the ten-dimensional coadjoint orbits of $SU(4)$. (The D–H polynomials for these orbits, unlike those for the six- and eight-dimensional orbits of $SU(4)$ or the six-dimensional orbits of $SU(3)$, don't seem to be easy to compute using just the Heckman formula alone.) Recall from Chapter 2 that the coadjoint orbits of $SU(4)$ are just the sets of "isospectral" 4×4 traceless Hermitian matrices. To be more specific, let α_i, $i = 1, \ldots, 4$, be a quadruple of real numbers, normalized so that $\alpha_1 \geq \alpha_2 \geq \alpha_3 \geq \alpha_4$ and $\alpha_1 + \cdots + \alpha_4 = 0$, and let $\mathcal{O}_\alpha$ be the set of all 4×4 Hermitian matrices whose eigenvalues are the α_is. Then every coadjoint orbit of $SU(4)$ is an $\mathcal{O}_\alpha$ and vice versa. The generic coadjoint orbits are those for which all the α_is are distinct, and these orbits are twelve-dimensional. The orbits in which we are interested at the moment are those for which two of the α_is are equal, and it turns out that these are exactly the ten-dimensional ones. Since the group G is the diagonal subgroup of $SU(4)$, $\mathfrak{g}^* \cong \mathfrak{g} \cong$ the space of 4×4 diagonal matrices of trace zero. The moment map $\Phi: \mathcal{O}_\alpha \to \mathfrak{g}^*$ is just the map that assigns to every matrix belonging to $\mathcal{O}_\alpha$ its diagonal entries. We will see in Chapter 5 that for the ten-dimensional orbits, its image Δ is either a "truncated tetrahedron" or a "skew-cubeoctahedron" depending on whether or not $\alpha_2 = \alpha_3$. (See Figures 3.11 and 3.12.)

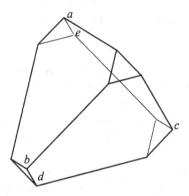

Fig. 3.11. Truncated tetrahedron.

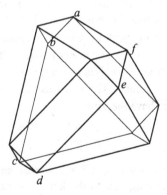

Fig. 3.12. Image of a 10-dimensional orbit: skew-cubeoctahedron.

The number of Δ_is in Figure 3.11 is 49 and in Figure 3.12 it is 15. (For a description of how the Δ_is sit inside the figures in each of these two cases, see Chapter 5.) We will consider the case $\alpha_2 \neq \alpha_3$ and, to be specific, we will assume $\alpha_1 = \alpha_2 = \alpha_0$, that is, that the matrices belonging to $\mathcal{O}_\alpha$ have the eigenvalues α_0, α_3, and α_4. As we mentioned previously, the walls of the Δ_is are the images, with respect to the moment map, of the fixed-point sets of certain circular subgroups of G. It turns out that, up to conjugation, there are only two subgroups that we have to worry about. The first is the group generated by the diagonal matrix with $+1$'s in its first two diagonal entries and -1's in the remaining two entries, and the second the group generated by the diagonal matrix with $+3$ in its first diagonal entry and -1's in the remaining diagonal entries. We will denote the generator of the first group by A and that of the second group by B. Consider now the action of each of these two groups of the coadjoint orbit $\mathcal{O}_\alpha$:

Proposition 3.5.3 *The fixed-point set of the group generated by A has four connected components, two $\mathbb{C}P^1$s and two $\mathbb{C}P^1 \times \mathbb{C}P^1$s, and for the $\mathbb{C}P^1 \times \mathbb{C}P^1$s the action of the quotient group G/S^1 is a Delzant action. The fixed-point set of the group generated by B has three components, two of which are $\mathbb{C}P^2$s; and the action of G/S^1 on these $\mathbb{C}P^2$s is a Delzant action.*

Proof. The fixed-point set of the group generated by A consists of the matrices in the set $\mathcal{O}_\alpha$ that commute with A; and these are the matrices that have the two-by-two block form

$$\begin{pmatrix} S & 0 \\ 0 & T \end{pmatrix}.$$

Thus the spectrum of S has to be either $\{\alpha_0\}$, $\{\alpha_3, \alpha_4\}$, $\{\alpha_0, \alpha_3\}$, or $\{\alpha_0, \alpha_4\}$, and the corresponding spectrum of T either $\{\alpha_3, \alpha_4\}$, $\{\alpha_0\}$, $\{\alpha_0, \alpha_4\}$, or $\{\alpha_0, \alpha_3\}$. It's easy to see that the set of matrices for which the spectrum of S is $\{\alpha_0\}$ or the spectrum of T is $\{\alpha_0\}$ is a $\mathbb{C}P^1$, and the set of matrices for which the spectrum of S is $\{\alpha_0, \alpha_3\}$ or the spectrum of T is $\{\alpha_0, \alpha_3\}$ is a $\mathbb{C}P^1 \times \mathbb{C}P^1$.

For B the situation is similar. The matrices belonging to the set $\mathcal{O}_\alpha$ that commute with B have to have the form:

$$\begin{pmatrix} s & 0 \\ 0 & T \end{pmatrix},$$

where $s = \alpha_0, \alpha_3$, or α_4 and T is a 3×3 Hermitian matrix with spectrum $\{\alpha_0, \alpha_3, \alpha_4\}$, $\{\alpha_0, \alpha_4\}$, or $\{\alpha_0, \alpha_4\}$. In the latter two cases the set of all such matrices is a $\mathbb{C}P^2$. □

Notice from Figure 3.12 that Δ has four exterior walls that are hexagonal in shape. From Proposition 3.5.3 it is easy to deduce that these are the only walls of the Δ_is that are not Delzant. In other words, we have:

Proposition 3.5.4 *For the ten-dimensional orbits of $SU(4)$, the jumps in the D–H polynomials across walls of the Δ_is are all of the form (3.90) except for the jumps across the four exterior hexagonal walls.*

In particular, at every interior wall, the change in f is a quadratic monomial whose level sets are planes parallel to the wall.

Appendix 3.A: The Duistermaat–Heckman Measure as the Volume of Reduced Spaces

Here is a proof of (3.10): Since G is abelian we can assume without loss of generality that $\alpha = 0$. Let $Y = \Phi^{-1}(0)$. Since the action of G on Y is free, Y is a principal G-bundle with M_0 as base; that is, there is a principal G-fibration

$$\pi: Y \to M_0. \tag{3.91}$$

Let's equip this principal bundle with a connection form v. (Recall that a connection form on (3.91) is a $\mathfrak{g}$-valued one-form whose restriction to the "typical fiber" G of (3.91) is the Maurer–Cartan form.) Consider on $\mathfrak{g}^* \times Y$ the two-form

$$d\langle pr_1, v\rangle + \pi^* \omega_0, \tag{3.92}$$

where ω_0 is the symplectic form on M_0. It is easy to check that (3.92) is a symplectic form in a sufficiently small neighborhood of $\{0\} \times Y$ and that the natural action of G on $\mathfrak{g}^* \times Y$ is Hamiltonian with respect to this form (with moment mapping pr_1). Moreover, there exists a G-invariant neighborhood of $\{0\} \times Y$ in $\mathfrak{g}^* \times Y$, a G-invariant neighborhood of Y in M, and a G-equivariant symplectomorphism of the first neighborhood into the second, mapping $\{0\} \times Y$ onto Y by the identity map. (This result is a special case of the "co-isotropic embedding" theorem; cf. [GS8], page 315.) Using this canonical form, let's compute the push-forward of the Liouville measure on M with respect to the moment mapping. Let $e_1, \ldots, e_k$ be a basis of $\mathfrak{g}$, let $x_1, \ldots, x_k$ be the coordinate functions on $\mathfrak{g}^*$ associated with this basis, and let

$$v = \sum v_i \otimes e_i,$$

the v_is being scalar one-forms on Y. Then, in terms of the x_is and v_is, (3.92) becomes

$$\sum dx_i \wedge v_i + \pi^* \omega_0 + \sum x_i dv_i.$$

On Y the last term vanishes; and, at points of Y, the Liouville form is

$$\pi^*(\omega_0^n) \wedge (\Pi v_i) \wedge dx, \tag{3.93}$$

n being half of the dimension of M_0. Suppose now that the push-forward of this form with respect to the moment map is $f(x)\, dx$. Then from (3.93) we get for $f(0)$ the expression

$$f(0) = \int \pi^*(\omega_0^n) \wedge \Pi v_i, \tag{3.94}$$

the integration being over Y. Recall now that since G is a compact, connected abelian group the exponential map exp: $\mathfrak{g} \longrightarrow G$ is a group epimorphism and its kernel is a sublattice Z_G of $\mathfrak{g}$. Let us choose the basis $e_2, \ldots, e_k$ to be an integral basis of Z_G. Then the k-form $\Pi \nu_i$ restricted to the typical fiber G of (3.91) is equal to the normalized G-invariant volume form on G; that is, its integral over G is 1. Hence the push-forward of the volume form

$$\pi^*(\omega_0^n) \wedge \Pi \nu_i$$

by π is just ω_0^n. In particular, the integral (3.94) is just the symplectic volume of M_0.

Appendix 3.B: Localization and the Duistermaat–Heckman Formula

We wish to prove the following formula: Let $G \times M \to M$ be a Hamiltonian action of a compact Lie group on a compact manifold M, with moment map $\Phi: M \to \mathfrak{g}^*$. Let $\xi \in \mathfrak{g}$ be an element of the Lie algebra $\mathfrak{g}$ of G whose corresponding vector field ξ_M has only isolated zeros. Let $M_0(\xi)$ denote the set of zeros of ξ_M; so $M_0(\xi)$ is a finite set. For each $p \in M_0(\xi)$ we have a linear transformation

$$L_p = L_p(\xi): TM_p \longrightarrow TM_p$$

given by

$$L_p(\xi)u = [\xi_M, \eta](p),$$

where η is any vector field with $\eta(p) = u \in TM_p$. The map L_p is invertible (otherwise p would not be an isolated zero) and has only imaginary eigenvalues. We can find an oriented basis $\{e_1, e_2, \ldots, e_{2n-1}, e_{2n}\}$ of TM_p such that

$$L_p e_{2j-1} = \lambda_j e_{2j},$$
$$L_p e_{2j} = -\lambda_j e_{2j-1},$$

where $\{\pm i\lambda_k\}$ are the eigenvalues of L_p. As explained in Section 3.2, the symplectic structure picks out the λ_j (as opposed to $-\lambda_j$) and hence we can define

$$(\det L_p)^{1/2} = \prod \lambda_j.$$

(Notice that this choice of the square root depends only on the orientation of TM_p and not on the symplectic structure.) Let

$$\mu = \omega^n/(2\pi)^n n!$$

denote the Liouville measure on M. Then the Duistermaat–Heckman formula is

$$\int_M e^{i\langle \Phi, \xi \rangle} \mu = i^n \sum_{p \in M_0(\xi)} \frac{e^{i\langle \Phi(p), \xi \rangle}}{(\det L_p)^{1/2}}. \tag{3.95}$$

The proof that we give here for this formula is taken from the book [BGV] and depends on the notion of equivariant differential forms. This subject is treated in detail in chapter VII of [BGV].

The first step is to move the i^n from the right-hand side of the equation to the left, and the $(2\pi)^n$ occurring in the denominator of μ from the left to the right, so that we can rewrite the Duistermaat–Heckman formula as

$$\int_M e^{i\langle \Phi, \xi \rangle} (i\omega)^n /n! = (-2\pi)^n \sum_{p \in M_0(\xi)} \frac{e^{i\langle \Phi(p), \xi \rangle}}{(\det L_p)^{1/2}}. \tag{3.96}$$

We can streamline this formula still further if we make the following convention: Let $\Omega = \Omega_{[0]} + \cdots + \Omega_{[2n]}$ be an inhomogeneous differential form, that is, a sum of differential forms of various degrees, $\Omega_{[k]}$ being the summand of homogeneous degree k. If N is an oriented k-dimensional submanifold of M, then we define

$$\int_N \Omega = \int_N \Omega_{[k]}.$$

For example, we can consider $\Omega = e^\omega = 1 + \omega + \frac{1}{2}\omega \wedge \omega + \cdots + \omega^n/n!$. Then the left-hand side of (3.96) can be written as

$$\int_M e^{i\langle \Phi, \xi \rangle} e^{i\omega}$$

or even more simply as

$$\int_M e^{i\omega(\xi)},$$

where we define the (inhomogeneous) form $\omega(\xi)$ as

$$\omega(\xi) = \omega + \langle \Phi, \xi \rangle.$$

On the other hand, if N is a point, then integration reduces to evaluation, and $\Omega(p)$ is defined to be $\Omega_{[0]}(p)$. So the numerator in the summand on the right-hand side of (3.95) and (3.96) can be written as

$$e^{i\omega(\xi)}(p).$$

So if we set $\alpha = e^{i\omega(\xi)}$ then (3.96) becomes

$$\int_M \alpha = (-2\pi)^n \sum_{p \in M_0(\xi)} \frac{\alpha(p)}{(\det L_p)^{1/2}}. \tag{3.97}$$

Define the operator d_ξ as

$$d_\xi \Omega = d\Omega - \iota(\xi_M)\Omega.$$

Notice that d_ξ is a derivation, so if $d_\xi \Omega = 0$ then $d_\xi f(\Omega) = 0$, where f is any polynomial or any entire function such as exp. Notice also that

$$d_\xi \omega(\xi) = 0$$

by the very definition of the moment map. Hence $d_\xi(e^{i\omega(\xi)}) = 0$. Thus the Duistermaat–Heckman formula is a special case of the following theorem.

Theorem 3.A.1 (Berline–Vergne localization theorem [BV]) *Suppose that G is a compact Lie group acting on a compact, oriented manifold M. Let $\xi \in g$ be such that the corresponding vector field ξ_M has isolated zeros, and let α be an inhomogeneous differential form satisfying*

$$d_\xi \alpha = 0.$$

Then (3.97) holds.

Proof. Let us first show that $\alpha_{[2n]}$ is exact outside $M_0(\xi)$. This would prove the theorem in the case that $M_0(\xi) = \emptyset$, the right-hand side of (3.97) being zero in this case. Put a G-invariant Riemannian metric on M, and define the one-form θ by

$$\theta(\eta) = (\xi_M, \eta).$$

Then $d_\xi \theta = d\theta - |\xi_M|^2$, and $d_\xi(d_\xi \theta) = 0$. Furthermore, the operator $\alpha \longmapsto d_\xi \theta \wedge \alpha$ is invertible outside of $M_0(\xi)$. Its inverse is given by $\alpha \longmapsto (d_\xi \theta)^{-1} \wedge \alpha$, where $(d_\xi \theta)^{-1}$ is a differential form satisfying

$$d_\xi(d_\xi \theta)^{-1} = 0.$$

If $d_\xi \alpha = 0$, then

$$\alpha = d_\xi[\theta(d_\xi \theta)^{-1}\alpha], \tag{3.98}$$

since d_ξ is a derivation. Taking the component of degree $2n$ of (3.98) shows that

$$\alpha_{[2n]} = d[\theta(d_\xi \theta)^{-1}\alpha]_{[2n-1]}.$$

Now let us examine what happens near a zero p of ξ_M. By the exponentiation map of the invariant metric, we can linearize ξ_M near p, so we can introduce coordinates $x_1, \ldots, x_{2n}$ so that

$$\xi_M = \lambda_1(x_2 \partial/\partial x_1 - x_1 \partial/\partial x_2) + \cdots + \lambda_n(x_{2n}\partial/\partial x_{2n-1} - x_{2n-1}\partial/\partial x_{2n})$$

near p. Define θ_p near p as

$$\theta_p = (1/\lambda_1)(x_2\, dx_1 - x_1\, dx_2) + \cdots + (1/\lambda_n)(x_{2n}\, dx_{2n-1} - x_{2n-1}\, dx_{2n}).$$

Then θ_p is just the form θ defined earlier, where we take the standard Riemannian metric

$$(1/\lambda_1^2)(dx_1 \otimes dx_1 + dx_2 \otimes dx_2) + \cdots + (1/\lambda_n^2)(dx_{2n-1} \otimes dx_{2n-1} + dx_{2n} \otimes dx_{2n}),$$

which is invariant under the action of the one-parameter group generated by ξ. Denote by T the closure of this subgroup in G. Then

$$\iota(\xi_M)\theta_p = x_1^2 + \cdots + x_{2n}^2 \overset{\text{def}}{=} \|x\|^2.$$

So we can patch these θ_p together with the θ defined earlier using a T-invariant partition of unity to get a θ that coincides with θ_p near p and such that $d_\xi \theta$ is invertible on $M - M_0(\xi)$. Thus by Stokes's theorem,

$$\int_M \alpha = -\lim_{\epsilon \to 0} \sum_{p \in M_0(\xi)} \int_{S_\epsilon} \theta(d_\xi \theta)^{-1}\alpha, \qquad (3.99)$$

where S_ϵ denotes the sphere $\|x\|^2 = \epsilon$ about p. Now the form $\theta = \theta + p$ is homogeneous of degree 2 in local coordinates, that is,

$$m_\epsilon^* \theta = \epsilon^2 \theta,$$

and so is $d_\xi \theta$. Hence

$$m_\epsilon^* [\theta(d_\xi \theta)^{-1}] = \theta(d_\xi \theta)^{-1}.$$

On the other hand, all the terms in $m_\epsilon^* \alpha$ tend to zero except the term of degree zero, which tends to $\alpha_{[0]}(p)$, which is defined to be $\alpha(p)$. Thus the limit of each summand in (3.99) is

$$\alpha(p) \int_{S_1} \theta(d_\xi \theta)^{-1}.$$

If we include the minus sign occurring in front of the sum, we have, on the unit sphere, where $\|x\|^2 = 1$,

$$-\theta(d_\xi \theta)^{-1} = \theta(1 - d\theta)^{-1} = \sum \theta(d\theta)^k.$$

As we are integrating over a sphere of dimension $2n - 1$, we must pick out the component of degree $2n - 1$, which is $\theta(d\theta)^{n-1}$. By Stokes's theorem,

$$\int_{S_1} \theta(d\theta)^{n-1} = \int_{B_1} (d\theta)^n,$$

where B_1 is the unit ball. But

$$(d\theta)^n = (-2)^n n! (\lambda_1 \cdots \lambda_n)^{-1} dx_1 \cdots dx_{2n}$$

by the binomial formula, and vol $B_1 = \pi^n/n!$. This completes the proof of the localization theorem.   □

4

Symplectic Fibrations and Multiplicity Diagrams

4.1 A Few Words about the Contents of This Chapter

The next few sections contain the main results of this monograph. In Section 4.2 we will discuss the following generalization of the Heckman formula of Section 3.3: Let G be an n-dimensional torus, let X and Y be compact symplectic manifolds, both equipped with Hamiltonian actions of G, and let

$$\pi: X \longrightarrow Y \qquad (4.1)$$

be a symplectic fibration that intertwines the two G actions. Then we will show that the Duistermaat–Heckman measure associated with the action of G on X is equal to the alternating sum

$$\sum (-1)^{w_p} \mu_p * v_p, \qquad (4.2)$$

where the ps are the fixed points of the action of G on Y, the v_ps are the v_ps of formula (3.50) of Chapter 3, and μ_p is the Duistermaat–Heckman measure associated with the action of G on the fiber above p. If the action of G on Y is fairly simple this formula can be a very effective computational tool. For example, suppose X is a generic coadjoint orbit of $U(n+1)$ and the subgroup G is the diagonal subgroup of $U(n+1)$. Using the results of Example 2.8 we will show in Section 4.3 that X fibers symplectically over $\mathbb{C}P^n$ and that the fibers over the fixed points are the coadjoint orbits of $U(n)$. In this case, (4.2) becomes an inductive formula for computing the D–H measure associated with a generic orbit of $SU(n+1)$ in terms of D–H measures associated with generic orbits of $U(n)$. (It is very closely related, by the way, to the Gelfand–Cetlin formula for computing the weight multiplicities of an irreducible representation of $SU(n+1)$. See, for instance, [Zel].)

Section 4.4 will be concerned with a "quantized" version of the formula (4.2); and to describe this formula we must first say a few words about how to "quantize" a symplectic fibration. Suppose to begin with that $\pi\colon X \to Y$ is a symplectic fibration in the sense of Section 1.1 (i.e., Y itself is not necessarily a symplectic manifold). Suppose that the symplectic form on each fiber F_p belongs to an integral cohomology class in $H^2(F_p, \mathbb{R})$. Suppose, in addition, that each fiber F_p is equipped with a polarization $\mathcal{F}_p$ that depends smoothly on p. Finally suppose that there exists a symplectic connection on X with the property that parallel transport along a curve γ in Y joining p to q carries the leaves of $\mathcal{F}_p$ into the leaves of $\mathcal{F}_q$. Let's now quantize each of the fibers F_p, following the standard geometric quantization procedure of Kostant–Souriau (see [K1], [K2]). This gives us for each F_p a Hilbert space E_p. We will denote by $E \to Y$ the vector bundle whose fiber at p is E_p. The symplectic connection on X gives rise to a connection on the vector bundle, which we'll denote by ∇_E.

Next let's suppose that Y itself is a symplectic manifold and that its symplectic form ω_B is integral. Then Y can be "prequantized" in the sense that there exists a Hermitian line bundle $\mathbf{L} \to Y$, and a connection $\nabla_\mathbf{L}$ on $\mathbf{L}$ such that

$$curv(\nabla_\mathbf{L}) = \omega_B.$$

The connections ∇_E and $\nabla_\mathbf{L}$ combine to give a connection on the tensor product, $E \otimes \mathbf{L}$, which we will denote simply by ∇.

Suppose finally that Y is also equipped with a polarization. Suppose, moreover, that this polarization has the following property: Let L be any leaf, and let i_L be the inclusion map of L into Y. *Then the symplectic connection on the pullback*

$$i_L^* X \longrightarrow L$$

is locally trivial. (See Example 2.4.)

Given this hypothesis we can associate the following quantum object to the fibration $X \to Y$: the space of sections of $\mathbf{L} \otimes E$ that are covariantly constant with respect to ∇ along the leaves of the foregoing polarization. Let's denote this space by $\mathcal{H}$.

The procedure we've just outlined is called "quantization in stages" (see [GLSW]). We will from now on make the additional assumptions that X and Y are compact and that the polarizations on the F_ps and on Y are positive definite. Then the vector bundle $E \to Y$ will be of finite rank and the Hilbert space $\mathcal{H}$ finite-dimensional.

Let's now suppose that X and Y are equipped with Hamiltonian actions of G that are intertwined by $\pi\colon X \to Y$. Moreover, let's suppose that the action of G on X preserves the $\mathcal{F}_p$s and the action on G or Y preserves the given foliation on Y. Then there is a natural representation of G on $\mathcal{H}$. In Section 4.4

we will prove the following "Kostant" theorem: Let $N(\alpha)$ be the multiplicity with which the weight $\alpha \in \mathfrak{g}^*$ occurs in the representation of G on $\mathcal{H}$. Then

$$N(\alpha) = \sum(-1)^{w_p}(N_p^\# * W_p)(\alpha). \tag{4.3}$$

Here the ps are the fixed points of the action of G on Y and the $N_p^\#$s the partition functions occurring in formula (3.59), that is,

$$N_p^\#(\alpha) = N_p(\alpha + \delta_p^w - (\Phi(p) + \delta)).$$

The W_ps are the multiplicity functions associated with the representation of G on the E_ps (they correspond to the μ_ps in formula (4.2)), and the convolution product is this discrete convolution product:

$$(N_p^\# * W_p)(\alpha) = \sum_{\alpha=\beta+\gamma} N_p^\#(\beta)W_p(\gamma) \tag{4.4}$$

summed over all weights β and γ with $\beta + \gamma = q$.

It will turn out, by the way, to be a good deal easier to prove (4.4) than it was to state it. We will see in Section 4.4 that (4.4) is a fairly easy consequence of the "Kostant" formula of Section 3.4.

In Section 4.5 we will look at what happens to (4.2) in the weak coupling limit. We will see that *lacunae* develop as the coupling parameter ϵ tends to zero. In fact the smaller ϵ becomes, the more the multiplicity diagram gets filled up by lacunary regions (until, at $\epsilon = 0$, it consists *entirely* of lacunary regions). On these regions, we will get from (4.2) a very simple formula for the D–H polynomials in terms of the D–H polynomials associated with Y.

(A parenthetical remark: Most weak coupling results require that the parameter ϵ be very small; however, the weak coupling results just described frequently turn out to be valid when ϵ is quite large. This is not too surprising given the fact that our data are piecewise polynomial functions of ϵ (see Chapter 5). For the $SU(4)$ results, which we will discuss in the next chapter, we will see that *all* lacunary phenomena can be accounted for by the weak coupling theorems of this section.)

In the remaining two sections of this chapter we will address a problem that we touched on briefly in the introduction to Chapter 3: Let M_α be the *reduced space* associated with a point α in our multiplicity diagram (Examples 3.7–3.9). Then the "multiplicity" attached to α can frequently be expressed as a symplectic invariant (such as the symplectic volume or Riemann–Roch number) of M_α. Therefore, M_α itself is an object of considerable interest in this theory, and it would be nice to know what the M_αs are for generic values of α. In particular it would be nice to know what the M_αs are for some of the simple examples that we looked at in Chapter 2. For instance, let M be a coadjoint orbit of $SU(n)$ and ψ the natural action of the Cartan subgroup G of $SU(n)$ on M. Let $\Phi: M \to \mathfrak{g}^*$ be the moment map associated with τ. Then, for a regular value α of Φ, it is known that M_α is a genuine symplectic manifold (that is, it is not just a symplectic "orbifold"). Moreover, it has a natural complex structure that is compatible with its symplectic structure (i.e., it is a *Kähler* manifold); and if α is integral it is even a projective variety. (See [GS8].)

However, though these properties make M_α an appealing object to study, its symplectic structure is still a mystery. What *is* known about the M_αs from the symplectic perspective can be summarized in a few lines:

1. As we pointed out in Example 3.8, for αs in a fixed Δ_i, the M_αs are diffeomorphic, and their symplectic structures vary in a "linear" fashion. (See formula (3.12).)

2. For regular values of α in different Δ_is the M_αs are "birationally equivalent." As α passes through a wall separating two adjacent Δ_is the symplectotype of M_α changes by a "blowing-up" followed by a "blowing-down." We won't say anything more about this (rather subtle) result in this monograph except to point out that the "blowing-up" referred to here is *symplectic* blowing-up in the sense of Gromov (see [McD]) and is quite different from blowing-up in the context of algebraic geometry. For more on this result see [GS9].

3. If α lies in a lacunary region or in a region abutting an exterior wall one can show that M_α fibers symplectically over a lower-dimensional symplectic manifold. In certain cases enough is known about the nature of this fibration to determine M_α up to symplectomorphism. (In fact, in the $SU(n)$ case, we will see that for the most interesting lacunary region, that containing the origin, the base of this fibration is a point!)

The purpose of the last two sections of this chapter will be to discuss item 3 in more detail. In Section 4.6 we will prove a result which says, roughly speaking, that the operation of reduction *commutes* with symplectic fibrations. This will explain why, for α in a lacunary region, M_α fibers over a lower-dimensional symplectic manifold. The fact that this is true when α is in a region contiguous to an exterior face requires a little more explanation: Let ξ be the vector in $\mathfrak{g}$ perpendicular to the hyperplane containing this exterior face, let S^1 be the subgroup of G generated by ξ, and let $f: M \longrightarrow \mathbb{R}$ be the moment map associated with the Hamiltonian action of S^1 on M. Without loss of generality we can assume that zero is a minimum value of f. Then

$$M_0 = \{p \in M \mid \quad f(p) = 0\}$$

is a connected symplectic submanifold of M on which S^1 acts trivially. Let's think of M for the moment as being a Hamiltonian S^1-space, and, for small positive values of ϵ, let

$$M_\epsilon = f^{-1}(\epsilon)/S^1$$

be the corresponding reduced space. One can show (using Theorem 2.2.1 of Chapter 2) that there is a symplectic fibration, with fiber $\mathbb{C}P^N$,

$$\pi\colon M_\epsilon \longrightarrow M_0 \qquad\qquad (4.5)$$

that commutes with the natural action of G/S^1 on M_ϵ and M_0. One can, therefore, get information about the M_αs, with α lying on the hyperplane $\langle \xi, \alpha \rangle = \epsilon$, by applying the weak coupling result proved in Section 4.6 to the fibration (2.3). (One can in fact show that these M_αs are symplectic fiber bundles with $\mathbb{C}^N$ as typical fiber, $2(N+1)$ being the codimension of M_0 in M.) Details can be found in Section 4.7.

4.2 The Heckman Formula

We'll begin our discussion of this formula by describing a kind of "baby version" of it: Let X be a compact Hamiltonian G-space and let $\Phi\colon X \to \mathfrak{g}^*$ be its moment map. The convexity theorem (Example 3.7) says that the image of Φ is the convex hull of the images of the fixed points of G; that is, if $q_1, \ldots, q_N$ are the fixed points of G, and $w_i = \Phi(q_i)$, the image of Φ is the convex polytope

$$\{s_1 w_1 + \cdots + s_N w_N \mid s_i \geq 0, \quad s_1 + \cdots + s_N = 1\}. \qquad (4.6)$$

Suppose now that X fibers symplectically over Y as in (4.1). Then it's clear that the fixed points of G in X have to sit on *fixed* fibers, that is, fibers F_p for which p is a fixed point of G in Y. Consider now the restriction of Φ to F_p. The image of $\Phi\colon F_p \to \mathfrak{g}^*$ is, by the convexity theorem, a convex polytope Δ_p, and by our previous remark the set (4.6) is identical with the set:

$$\left\{ \sum_p t_p w_p \;\middle|\; t_p \geq 0, \; \sum_p t_p = 1, \; w_p \in \Delta_p \right\}. \qquad (4.7)$$

In other words, the image of Φ is the convex hull of the Δ_ps. This trivial observation turns out to be very useful for seeing what the images of moment maps look like. For instance, we will see shortly that the generic coadjoint orbit of $U(n)$ fibers symplectically over $\mathbb{C}P^{n-1}$. If G is the Cartan subgroup of $U(n)$, then, acting on $\mathbb{C}P^{n-1}$, it has n fixed points; and their images in $\mathfrak{g}^*$ (i.e., in $\mathbb{R}^n$) are the points $(1, 0, \ldots 0), (0, 1, \ldots, 0), \ldots$. Thus the image of $\mathbb{C}P^{n-1}$ is the standard $(n-1)$-simplex

$$\{(x_1, \cdots, x_n) \mid x_i \geq 0, \; x_1 + \cdots + x_n = 1\}. \qquad (4.8)$$

The fibers over the fixed points are generic coadjoint orbits of $U(n-1)$, so the preceding remark gives us an inductive method for computing the images of the generic $U(n)$ orbits. To illustrate the efficacy of this method let's actually carry out the first few steps in the induction: Let's begin with the generic orbit of $U(2)$. This is just $\mathbb{C}P^1$, and, as we pointed out earlier, its image is the one-simplex,

$$\{t_1 + t_2 = 1 \mid \quad t_1, t_2 \geq 0\}, \tag{4.9}$$

that is, it is just an interval. Next consider a generic orbit $\mathcal{W}_3$ of $U(3)$. This orbit fibers symplectically over $\mathbb{C}P^2$, whose image is, by (4.8), an equilateral triangle. (See Figure 4.1.) To compute the image of $\mathcal{W}_3$ we must replace the

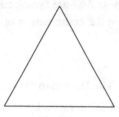

Fig. 4.1.

vertices of this triangle by the Δ_ps of (4.7); but, by (4.9), these are just intervals. We will show in the next section that the interval that replaces a given vertex has to be parallel to the opposite side, and, by Weyl group symmetry, these intervals have to be of the same length. (See Figure 4.2.) Hence, connecting the dots, we

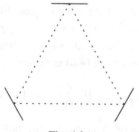

Fig. 4.2.

get for the image of a generic orbit of $U(3)$ the hexagon in Figure 4.3. Let's carry this induction one stage further and compute the image of a generic orbit of $U(4)$. By (4.8) the image of $\mathbb{C}P^3$ is a regular tetrahedron (see Figure 4.4) and by induction the images of the Δ_ps are hexagons. Again we will see in the next section that the hexagon that replaces a given vertex has to be parallel to the opposite face and by Weyl group symmetry all these hexagons have to be

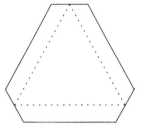

Fig. 4.3.

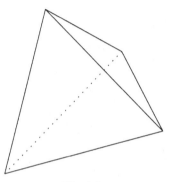

Fig. 4.4.

congruent. (See Figure 4.5.) Hence, connecting the dots, we get for the image of a generic orbit of $U(4)$ the solid pictured in Figure 4.6.

Let's now turn to the formula (4.2) itself. For simplicity we will assume that the action of G on X has finitely many fixed points. Then by formula (3.50) of Chapter 3 the D–H measure associated with the action of G on X is a sum over the fixed points of the terms

$$(-1)^{w_q}(L_q)_* ds, \tag{4.10}$$

where ds is Lebesgue measure on the positive orthant in $\mathbb{R}^N$:

$$\{(s_1, \cdots, s_N) \mid s_i \geq 0\},$$

and L_q is the linear mapping

$$L_q(s) = \sum_{i=1}^N s_i \alpha_{q,i}^w + \Phi(q). \tag{4.11}$$

The $\alpha_{q,i}$s in (4.11) are the weights of the isotropy representation of G on the tangent space to X at q and the $\alpha_{q,i}^w$s are the renormalized weights relative to

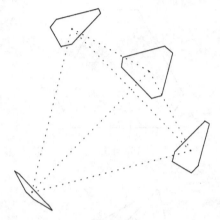

Fig. 4.5.

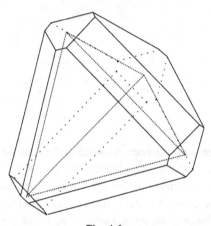

Fig. 4.6.

a choice of "positive Weyl chamber." (See formula (3.50) and the discussion preceding it.)

Let's now come back to the observation we made earlier: The fixed points have to sit on fibers F_p over fixed points p of G in Y. Let ξ be an element of $\mathfrak{g}$ and $\xi^\#$ the vector field on X associated with it. Then

$$d\langle \Phi, \xi \rangle = \iota(\xi^\#)\omega_X = \iota(\xi^\#)\pi^*\omega_Y + \iota(\xi^\#)\omega_\Gamma, \qquad (4.12)$$

where ω_X and ω_Y are the symplectic forms on X and Y and ω_Γ the minimal coupling form.

For the moment, fix p and let $F = F_p$. Let $\phi_F\colon F \to \mathfrak{g}^*$ and $\phi_Y\colon Y \to \mathfrak{g}^*$ be the moment maps associated with the action of G on F and on Y. Along F, $\xi^\#$

is tangent to F; so the first term in (4.12) is zero, and the second term is equal to $\langle d\phi_F, \xi \rangle$. Thus this equation becomes

$$\langle d\Phi, \xi \rangle = \langle d\phi_F, \xi \rangle \qquad \text{for all } \xi \in \mathfrak{g}. \tag{4.13}$$

Notice that this equation only determines ϕ_F up to an additive constant. However, (4.12) suggests that the correct choice of this constant ought to be

$$\phi_F(q) + \phi_Y(p) = \Phi(q) \tag{4.14}$$

for all points q above p; in particular for the fixed point q of equation (4.11).

From now on we will assume that (4.14) holds at all fixed points q. Coming back to (4.11), we see that the set of weights $\alpha_{q,i}$ consists of weights associated with the representation of G on the *tangent* space to F at q and weights associated with the representation of G on the *normal* space to F at q. By relabeling we can assume that the weights $\alpha_{q,i}, i = 1, \ldots, K$, are weights of the first kind and the weights $\alpha_{q,i}, i = K + 1, \ldots, N$, are weights of the second kind. Let $l = N - K$, and let's relabel these weights

$$\beta_{p,i} = \alpha_{q,K+i}, \qquad i = 1, \ldots, l. \tag{4.15}$$

Notice that we are justified in replacing the suffix q by a suffix p in (4.15) because the $\beta_{p,i}$s are the weights of the representation of G on the tangent space to Y at p.

Setting $t_i = s_{K+i}$, we can, in view of (4.14) and (4.15), rewrite (4.11):

$$L_q(s,t) = \left(\sum_{i=1}^{K} s_i \alpha_{q,i}^w + \phi_F(q) \right) + \left(\sum_{j=1}^{l} t_j \beta_{p,j}^w + \phi_Y(p) \right). \tag{4.16}$$

Now let's apply Theorem 3.3.3 to the symplectic manifold F. Theorem 3.3.3 says that the D–H measure μ_F associated with the action of G on F is the alternating sum over the fixed points q of G on F of the measures

$$(L_{F,q})_*(ds_1 \cdots ds_K),$$

where $L_{F,q}$ is the map

$$(s_1, \ldots, s_K) \longrightarrow \sum_{i=1}^{K} s_i \alpha_{q,i}^w + \phi_F(q).$$

On the other hand, the D–H measure associated with the action of G on Y is the sum

$$\sum (-1)^{w_p} v_p,$$

where v_p is the measure

$$(L_{Y,p})_*(dt_1 \cdots dt_l)$$

and $L_{Y,p}$ the map

$$(t_1, \cdots, t_l) \longrightarrow \sum_{i=1}^{l} t_i \beta_{p,i}^w + \Phi_Y(p).$$

However, by (4.16)

$$L_q(s, t) = (L_{F,q})(s) + (L_{Y,p})(t),$$

so

$$(L_q)_* ds\, dt \ = \ (L_{F,q})_* ds * (L_{Y,p})_* dt.$$

The second term in this convolution product is v_p, so, with p fixed, the alternating sum of these measures over the fixed points on F is $\mu_F * v_p$. Multiplying this by $(-1)^{w_p}$ and summing over p we obtain (4.2).

4.3 An Inductive Formula for the D–H Measure Associated with a Coadjoint Orbit of $U(n)$

We will make the usual identification

$$A \ \longrightarrow \ \text{trace}(A\cdot)$$

of $\mathfrak{u}(n)$ with $\mathfrak{u}(n)^*$, and we will also make the usual identification of $\mathfrak{u}(n)$ with $\sqrt{-1}\,\mathfrak{u}(n)$, the space of Hermitian $n \times n$ matrices. Via these identifications coadjoint orbits of $U(n)$ become *isospectral* sets of Hermitian matrices. More explicitly, let

$$\lambda_1 \geq \lambda_2 \geq \cdots \geq \lambda_n \qquad\qquad (4.17)$$

be a sequence of real numbers, and let $\mathcal{W}_\lambda$ be the set of all Hermitian matrices A, with $\text{spec}(A) = (\lambda_1, \ldots, \lambda_n)$. Then, via these identifications the map $\lambda \to \mathcal{W}_\lambda$ sets up a one-to-one correspondence between sequences of the form (4.17) and coadjoint orbits of $U(n)$.

Now let λ be such a sequence and suppose that one of the λ_is in this sequence is distinct from all the others. For simplicity let's suppose that this is λ_n (i.e., let's suppose that λ_n is strictly less than the other λ_is). This implies that if $A \in \mathcal{W}_\lambda$, the eigenspace of A associated with λ_n is one-dimensional. This one-dimensional subspace of $\mathbb{C}^n$ corresponds to a *point* in $\mathbb{C}P^{n-1}$. Let's denote this point by $\pi(A)$. It follows from Corollary 2.3.5 that the map

$$\pi \colon \mathcal{W}_\lambda \ \longrightarrow \ \mathbb{C}P^{n-1} \qquad\qquad (4.18)$$

sending A to $\pi(A)$ is a symplectic fibration.

Now let T^n be the diagonal subgroup of $U(n)$ and let's see what the formula (4.2) says when π is the fibration (4.18). The fixed points of the action of T^n on $\mathbb{C}P^{n-1}$ are the points corresponding to the one-dimensional subspaces $\{ce_i\}$, where e_i is the ith standard basis vector of $\mathbb{C}^n$. Let's denote these fixed points by $p_1, \ldots, p_n$. The stabilizer group of p_i in $U(n)$ consists of all matrices $A \in U(n)$ with $Ae_i = ce_i$ for some complex number c of modulus one, and it is isomorphic to $U(n-1) \times S^1$.

In particular, it contains inside it a copy of $U(n-1)$. Let F_i be the fiber of $\mathcal{W}_\lambda$ above p_i. This fiber consists of all $A \in \mathcal{W}_\lambda$ with $Ae_i = \lambda_n e_i$; so it is clear that $U(n-1)$ acts *transitively* on it. In fact, F_i can be canonically identified with the (co)adjoint orbit $\mathcal{W}_{\lambda'}$ of $U(n-1)$, where

$$\lambda' = (\lambda_1, \ldots, \lambda_{n-1}).$$

The group T^n acts on this fiber in the following fashion: The subgroup

$$\{(1, 1, \ldots, e^{\sqrt{-1}\theta_i}, \ldots, 1)\}$$

acts trivially, and the complementary subgroup

$$T^{n-1} = \{(e^{\sqrt{-1}\theta_i}, \ldots, e^{\sqrt{-1}\theta_{i-1}}, 1, e^{\sqrt{-1}\theta_{i+1}}, \ldots, e^{\sqrt{-1}\theta_n})\} \qquad (4.19)$$

can be identified with the diagonal subgroup of $U(n-1)$; and, if we make this identification and identify F_i with $\mathcal{W}_{\lambda'}$, its action is the (co)adjoint action. In particular, let $m_{\lambda'}$ be the D–H measure associated with the action of $U(n-1)$ on $\mathcal{W}_{\lambda'}$ and let

$$\delta_i \colon \mathbb{R}^{n-1} \longrightarrow \mathbb{R}^n \qquad (4.20)$$

be the mapping

$$\delta_i(y_1, \ldots, y_{n-1}) = (y_1, \ldots, y_{i-1}, 0, y_i, \ldots, y_{n-1}).$$

If we identify the Lie algebra of T^{n-1} (respectively, T^n) with $\mathbb{R}^{n-1}$ (respectively, $\mathbb{R}^n$) then (4.20) is just the morphism of Lie algebras corresponding to the embedding of T^{n-1} into T^n given by (4.19). Thus *the D–H measure associated with the action of T^n on F_i is just*

$$(\delta_i)_* m_{\lambda'}. \qquad (4.21)$$

Finally let's compute the measures v_i associated with the fixed points q_i. Let V_i be the one-dimensional space $\{ce_i\}$. Then the tangent space to $\mathbb{C}P^{n-1}$ at p_i is

$$(\mathbb{C}^n / V_i) \otimes V_i^*$$

and the weights of the action of T^n on this space are

$$\alpha_{i,j} = \theta_j - \theta_i, \quad j = 1, \ldots, i-1, i+1, \ldots, n. \tag{4.22}$$

If we make the standard choice of a positive Weyl chamber (the set consisting of positive linear combinations of the $\alpha_{j,j+1}$s) then

$$\alpha_{i,j}^w = \begin{cases} \alpha_{i,j} & \text{if } i < j, \\ -\alpha_{i,j} & \text{if } j < i. \end{cases} \tag{4.23}$$

In particular, $(-1)^{w_i} = (-1)^{i-1}$. Consider now the set

$$\left\{ \sum_{j \neq i} s_j \alpha_{i,j}^w \;\middle|\; s_j \geq 0 \right\}. \tag{4.24}$$

Letting

$$\sum_{j \neq i} s_j \alpha_{i,j}^w = \sum_{j=1}^n x_j \theta_j$$

we get from (4.22) and (4.23)

$$\begin{aligned} x_j &= \quad s_j, \quad j > i, \\ x_j &= \quad -s_j, \quad j < i, \\ x_i &= -\textstyle\sum_{k>i} s_k + \sum_{k<i} s_k. \end{aligned} \tag{4.25}$$

Thus, in the x-coordinates, the set (4.24) is defined by the (in)equalities

$$\sum x_j = 0, \quad x_j \geq 0 \quad \text{for} \quad j > i, \quad x_j \leq 0 \quad \text{for} \quad j < i. \tag{4.26}$$

The mapping

$$L_i(s) = \sum_{j \neq i} s_j \alpha_{i,j}^w$$

maps the positive orthant in s-space *bijectively* onto this set. *Thus, up to a constant factor, v_i is standard Lebesgue measure on the set* (4.26).

Putting together these comments, and recalling the formula (4.21), we get from (4.2) the following formula for the D–H measure associated with the coadjoint orbit $\mathcal{W}_\lambda$:

$$m_\lambda = \sum_{i=1}^n (-1)^{i-1} v_i * (\delta_i)_* m_{\lambda'}. \tag{4.27}$$

In other words, the measure m_λ is obtained from the measure $m_{\lambda'}$ by applying to it the "coboundary operation":

$$\sum (-1)^{i-1} v_i * (\delta_i)_*. \tag{4.28}$$

Notice, by the way, that if $\lambda = (1, \ldots, 1, 0)$, then $W_\lambda = \mathbb{C}P^{n-1}$, $W_{\lambda'} = \mathbb{C}P^{n-2}$, and the formula (4.27) is just the "simplicial form" of Stokes's theorem (which says that if one applies the coboundary operation (4.28) to Lebesgue measure on the standard $(n-2)$-simplex one gets Lebesgue measure on the standard $(n-1)$-simplex).

A final comment is useful concerning this formula. Let Σ_n be the group of permutations of $(1, \ldots, n)$. This group acts, by coordinate permutations, on the space $\mathbb{R}^n$, and this action preserves the measure m_λ. It doesn't, however, preserve the individual summands on the right-hand side of (4.27), so for every $\sigma \in \Sigma_n$ we get a *new* formula for m_λ, namely,

$$m_\lambda = \sum (-1)^{i-1} (v_i * (\delta_i)_* m_{\lambda'})^\sigma, \qquad (4.29)$$

where $(v_i * (\delta_i)_* m_{\lambda'})^\sigma$ is the measure obtained by applying the linear transformation

$$(x_1, \ldots, x_n) \longrightarrow (x_{\sigma(1)}, \ldots, x_{\sigma(n)})$$

to the measure $v_i * (\delta_i)_* m_{\lambda'}$. We leave the following as an exercise.

Proposition 4.3.1 *For every point x in the support of m_λ there exists a $\sigma \in \Sigma_n$ with the property that all summands but one on the right-hand side of (4.29) are zero in a neighborhood of x.*

Thus, *collectively*, the formulas (4.28) provide a very efficient inductive scheme for computing the measures m_λ. (It would be interesting to see if, for $U(4)$, this could be implemented on a computer!)

4.4 The Kostant Formula

The proof we will give here is quite similar to the proof we just gave of (4.2). By the Kostant formula of Section 3.4, the weight multiplicities of the representation associated with X are given by an alternating sum over the fixed points q of G of a partition function $N_q^{\#}$.

Just as in Section 4.2 we can write $N_q^{\#}$ as the convolution product of a fiber term and a base term. Summing over all qs on a fixed fiber F_p, we obtain, by the Kostant formula for F_p, the weight multiplicities of the representation of G on E_p. If we now sum over the ps we get (4.3). Here are the details:

Let q be a fixed point of G lying on the fixed fiber $F = F_p$. By formula (3.59) the contribution of q to the multiplicity formula for X is

$$(-1)^{w_q} N_q^{\#}, \qquad (4.30)$$

where

$$N_q^{\#}(\nu) = N_q(\nu - \Phi(q) + \delta_q^w - \delta_q). \tag{4.31}$$

We'll remind our readers how the individual terms in this formula are defined: As in Section 4.2 let $\alpha_{q,i}$, $i = 1, \ldots, k$, be the weights of the representation of G on the tangent space to F at q and $\beta_{p,j}$, $j = 1, \ldots, l$, the weights of the representation of G on the tangent space to Y at p. Then the αs and βs combined are the weights of the representation of G on the tangent space to X at q. Now fix a positive Weyl chamber and replace the αs and βs by the "renormalized" αs and βs, that is, $\alpha_{q,i}^w$, $i = 1, \ldots, k$, and $\beta_{p,j}^w$, $j = 1, \ldots l$. (Recall that "renormalizing" means fixing some element $\xi \in \mathfrak{g}$ that sits in the interior of the "positive Weyl chamber" and replacing $\alpha_{q,i}$ (respectively, $\beta_{p,j}$) by $-\alpha_{q,i}$ (respectively, $-\beta_{p,j}$) if $\alpha_{q,i}(\xi)$ (respectively, $\beta_{p,j}(\xi)$) is negative.) The number w_p is the number of αs and βs that get changed into their negatives by renormalization and δ_p and δ_p^w are the sums

$$\frac{1}{2}\left(\sum \alpha_{q,i} + \sum \beta_{p,j}\right) \quad \text{and} \quad \frac{1}{2}\left(\sum \alpha_{q,i}^w + \sum \beta_{p,j}^w\right). \tag{4.32}$$

Finally N_p is the Kostant partition function: Evaluated on a weight ν, $N_p(\nu)$ is the number of ways that one can write ν as a sum

$$\nu = \sum m_i \alpha_{q,i}^w + n_j \beta_{p,j}^w, \tag{4.33}$$

the m_is and n_js being nonnegative integers.

We will now decompose all the objects we've just described into their "fiber" and "base" components (using superscript "primes" to denote fiber components and replacing subscript qs by subscript ps to denote base components). To begin with let $N_q'(\alpha)$ be the number of ways that one can write the weight α as a sum

$$\alpha = \sum m_i \alpha_{q,u}^w,$$

the m_is being nonnegative integers; and, similarly, let $N_p(\beta)$ be the number of ways that one can write β as a sum

$$\beta = \sum n_j \beta_{p,j}^w,$$

the n_js being nonnegative integers. It's clear from (4.33) that

$$N_q(\nu) = \sum_{\alpha+\beta=\nu} N_q'(\alpha) N_p(\beta), \tag{4.34}$$

the sum on the right being over all weights α and β with $\alpha + \beta = \nu$. Next notice that

$$\delta_q = \delta_q' + \delta_p \quad \text{and} \quad \delta_q^w = (\delta_q')^w + \delta_p^w, \tag{4.35}$$

where

$$\delta'_q = \frac{1}{2} \sum \alpha_{q,i} \quad \text{and} \quad (\delta'_q)^w = \frac{1}{2} \sum \alpha^w_{q,i}$$

and δ_p and δ^w_p are the same expressions with the $\alpha_{q,i}$s replaced by the $\beta_{p,j}$s. Moreover, by (4.14) (with the preceding notational conventions)

$$\Phi(a) = \Phi'(q) + \Phi(p). \tag{4.36}$$

Thus, for every decomposition of ν into a sum of weights, $\nu = \alpha + \beta$, we can rewrite (4.31) in the form

$$N^\#_q(\nu) = N_q(\alpha - \phi'(q) + (\delta'_q)^w - \delta'_q + \beta - \Phi(p) + \delta^w_p - \delta_p), \tag{4.37}$$

whence, by (4.34), we get for $N^\#_q(\nu)$ the sum

$$\begin{aligned} N^\#_q(\nu) &= \sum_{\alpha+\beta=\nu} (N'_q)^\#(\alpha) N^\#_p(\beta) \\ &\overset{\text{def}}{=} \quad \left((N'_q)^\# * N^\#_p\right)(\nu), \end{aligned} \tag{4.38}$$

where

$$(N'_q)^\#(\alpha) = N'_q(\alpha - \Phi'(q) + (\delta'_q)^w - \delta'_q) \tag{4.39}$$

and

$$N^\#_p(\beta) = N_p(\beta - \Phi(p) + \delta^w_p - \delta_p). \tag{4.40}$$

Notice finally that $w_q = w'_q + w_p$, where w'_q is the number of $\alpha_{q,i}$s that get changed into their negatives by renormalization and w_p the number of $\beta_{p,j}$s that get changed this way.

Thus the contribution (4.33) of q to the multiplicity formula can be written in this "base–fiber" notation:

$$(-1)^{w_p}\left((-1)^{w_i}(N'_q)^\#\right) * N^\#_p.$$

Summing over all fixed points q on the fiber $F = F_p$, we obtain

$$(-1)^{w_p} W_p * N_p,$$

where

$$W_p = \sum_{q \in F} (-1)^{w'_q} (N'_q)^\#. \tag{4.41}$$

This, however, is just the Kostant formula of Section 3.4 for the multiplicity function associated with the symplectic manifold F. Hence if the vector space E_p is the "quantum object" associated with F, it counts the weight multiplicities of the representation of G on E_p. Summing over the fixed points p of G in Y we obtain the formula (4.3).

4.5 The Weak Coupling Limit

Let ω_X and ω_Y be the symplectic forms on X and Y, respectively. As we pointed out in Chapter 1, ω_X defines a symplectic connection on the fiber bundle $\pi: X \to Y$. Moreover,

$$\omega_X = \omega_\Gamma + \pi^* \omega_Y, \qquad (4.42)$$

ω_Γ being the "minimal coupling form" associated with the connection. Let's now replace ω_X by the symplectic form

$$(\omega_X)_\epsilon = \epsilon \omega_\Gamma + \pi^* \omega_Y. \qquad (4.43)$$

What we want to do in this section is examine how (4.2) behaves when we rescale ω_X this way and let $\epsilon \to 0$. To begin with, let's look at the extreme situation: Y is equal to a point. Let $2N$ be the dimension of X. Then if we multiply ω_X by ϵ the Liouville form ω_X^N gets rescaled by the factor ϵ^N and the moment map by the factor ϵ; so if μ is the D–H measure associated with the original symplectic form and μ_ϵ the rescaled form, then

$$\mu_\epsilon = \epsilon^N (\lambda_\epsilon)_* \mu, \qquad (4.44)$$

where $\lambda_\epsilon: \mathfrak{g}^* \longrightarrow \mathfrak{g}^*$ is the homothety $\nu \longrightarrow \epsilon \nu$. In particular, as ϵ tends to zero, $\epsilon^{-N} \mu_\epsilon$ tends weakly to $C\delta_0$, C being the original symplectic volume of X. This convergence, in fact, is a little bit better than just weak convergence:

If Δ is the support of μ, the support of μ_ϵ is $\epsilon \Delta$, so the *support* of μ_ϵ converges to the support of the delta function δ_0.

Now let $\pi: X \longrightarrow Y$ be the symplectic fibration (4.1) and μ and μ_ϵ the D–H measures associated with the symplectic forms (4.42) and (4.43), respectively. By (4.2) μ is equal to the sum over the fixed points p of G in Y:

$$\sum (-1)^{w_p} \mu_p * v_p, \qquad (4.45)$$

μ_p being the D–H measure corresponding to the Hamiltonian action of G on the fiber F_p. When we rescale, the v_ps, which depend only on the symplectic structure of Y, are unchanged. The μ_ps on the other hand rescale according to the formula (4.44) (except that we have to replace the "N" in (4.44) by the fiber dimension r of F_p). Thus

$$\mu_\epsilon = \epsilon^r \sum (-1)^{w_p} (\lambda_\epsilon)_* \mu_p * v_p. \qquad (4.46)$$

Notice that if we divide both sides by ϵ^r and let ϵ tend to zero, then the terms $(\lambda_\epsilon)_* \mu_p$ tend to δ_0, so

$$\epsilon^{-r} \mu_\epsilon \longrightarrow \sum (-1)^{w_p} v_p = v, \qquad (4.47)$$

the measure v being the D–H measure associated with the action of G on Y

(formula (3.50)). Let's now examine a little more carefully how the individual terms in the sum (4.47) tend to their respective limits.

In the notation of Section 4.2, the support of v_p is the cone

$$C_p = \left\{ \Phi(p) + \sum s_j \beta^w_{p,j} \;\middle|\; s_j \geq 0 \right\}, \tag{4.48}$$

and the support of $(\lambda_\epsilon)_* \mu_p$ is the convex polytope $\epsilon \Delta_p$. Thus the support of the convolution of these two measures $(\lambda_\epsilon)_* \mu_p * v_p$ is the set $\epsilon \Delta_p + C_p$. Let

$$\mu_p = f_p(s)ds \quad \text{and} \quad v_p = g_p(x)dx, \tag{4.49}$$

f_p and g_p being the Radon–Nikodym derivatives of μ_p and v_p with respect to ds and dx. Then

$$(\lambda_\epsilon)_* \mu_p * v_p = h_{p,\epsilon}(x)dx, \tag{4.50}$$

where

$$h_{p,\epsilon}(x) = \epsilon^{-n} \int f_p\left(\frac{y}{\epsilon}\right) g_p(x - y)\, dy. \tag{4.51}$$

The results we proved about linear symplectic actions in Section 3.2 tell us that we can subdivide C_p into convex conic subsets

$$C_p = \bigcup C_{p,i} \tag{4.52}$$

such that on each $C_{p,i}$ the function g_p is a polynomial of degree equal to dimension $Y/2-$ dimension G. Suppose now that x sits in the set

$$\bigcap (C_{p,i} + \epsilon y) \tag{4.53}$$

(the intersection being over all $y \in \Delta_p$). Then the integrand in (4.51) is a polynomial of the same degree in x, and so we've proved:

Theorem 4.5.1   *On the set (4.53) the function $h_{p,\epsilon}$ is a polynomial of degree equal to* (dim Y)/2 $-$ dim G.

In particular, (4.53) is a *lacunary* region for the measure, $(\lambda_\epsilon)_* \mu_p * v_p$, since the D–H polynomials associated with the measure are generically of degree equal to (dim X)/2 $-$ dim G. Notice also that as ϵ tends to zero the region (4.53) tends to $C_{p,i}$, so we've proved:

Theorem 4.5.2   *As $\epsilon \to 0$ the lacunary regions associated with the piecewise-polynomial measure $(\lambda_\epsilon)_* \mu_p * v_p$ fill up all of C_p.*

Summing over the ps we get the same assertion for the measure μ_ϵ itself: Namely, let Δ be the support of the measure v, and let

$$\bigcup \Delta_i$$

be the subdivision of Δ into polytopes Δ_i on which this measure is a polynomial multiple of Lebesgue measure (say $f_i(x)dx$). Then, as a corollary of the previous results, we obtain:

Theorem 4.5.3 *Let x_0 be an interior point of Δ_i. Then, for ϵ sufficiently small, there exists a neighborhood of x_0 on which $\mu_\epsilon = f_{i,\epsilon}(x)dx$, $f_{i,\epsilon}(x)$ being a polynomial of degree less than or equal to* dim $Y/2$ − dim G.

As an illustration of this result let's look again at the example discussed in Section 4.3. As we've pointed out already, the measure v associated with $\mathbb{C}P^{n-1}$ is Lebesgue measure on the standard $(n-1)$-simplex:

$$\{s_1 + \cdots + s_n = 1 \mid \quad s_1 \geq 0\}.$$

It is not hard to see that, as long as the λ_is in the sequence (4.17) are distinct, the barycenter of this simplex (the point $s_1 = \cdots = s_n = 1/n$) lies in a lacunary region on which the D–H polynomial is *constant*. For instance, if $n = 3$ and $\lambda = (\lambda_1, \lambda_2, \lambda_3)$ with $\lambda_1 > \lambda_2 > \lambda_3$, the support of m_λ, as we saw in Section 4.2, is a hexagon (Figure 4.3). The lacunary region in this hexagon is the shaded equilateral triangle in Figure 4.7. The weak-coupling limit for this

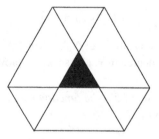

Fig. 4.7.

figure corresponds to setting $\lambda_1 = a + \epsilon$, $\lambda_2 = a - \epsilon$, and $\lambda_3 = b$ with $a > b$, and letting $\epsilon \to 0$.

In this limit two things happen: The hexagon gets deformed into the triangle with vertices (a, q, b), (a, b, q), and (b, a, q), and this triangle becomes identical with the shaded triangle in the figure.

4.6 Reduction and Weak Coupling

Let Φ and ψ be the moment mappings associated with the actions of G on X and Y. Then if α is a regular value both of ψ and of Φ we can form the reduced spaces

$$M_\alpha = \Phi^{-1}(\alpha)/G$$

and

$$N_\alpha = \psi^{-1}(\alpha)/G.$$

The main result of this section says, roughly speaking, that there exists a symplectic fibration

$$\pi_\alpha: M_\alpha \longrightarrow N_\alpha \tag{4.54}$$

whose typical fiber is the same as the typical fiber of the fibration (4.1). Strictly speaking we will only be able to prove this result in the weak coupling limit; however, as we remarked earlier, many of the weak coupling results of this chapter are true for "unreasonably large" values of the coupling parameter because of the piecewise-polynomial behavior of the objects in this theory. This also holds true for the results of this section.

For instance, we pointed out in Section 4.5 that for generic coadjoint orbits of $U(n)$ there exists a lacunary region containing the origin where the multiplicity function is constant. If one applies the results of this section to the fibration (4.1), one gets the following symplectic analog of this result: Let a and b be real numbers, with $a > b$, and let λ be a sequence of the form (4.17) that is "ϵ-close" to the sequence whose first $n-1$ terms are a and whose last term is b.

Since the action of T^n on $\mathbb{C}P^{n-1}$ is a "Delzant action" (Section 3.5) the base space N_α in (4.54) is just a point; so, for α near zero, M_α is symplectomorphic to the $U(n-1)$ coadjoint orbit $\mathcal{W}_{\lambda'}$, where $\lambda' = (\lambda_1, \ldots, \lambda_{n-1})$.

Let's now turn to the details. The mappings π, Φ, and ψ give rise to a diagram:

$$\begin{array}{ccc} & X & \\ {\scriptstyle\pi}\downarrow & & \searrow{\scriptstyle\Phi} \\ Y & \overset{\psi}{\longrightarrow} & \mathfrak{g}^* \end{array} \tag{4.55}$$

Unfortunately, this diagram doesn't commute. (If we knew that it did commute, exhibiting a fibration of the form (4.54) would be a relatively easy matter since we would have

$$\Phi^{-1}(\alpha) = \psi^{-1}(\pi^{-1}(\alpha)),$$

and hence, because π is G-equivariant, it would automatically induce a mapping of M_α into N_α.) What we will show, however, is that one can modify the symplectic structure of X (without changing its symplectotype) such that there exists a neighborhood U of $\psi^{-1}(\alpha)$ over which the diagram (4.55) *does* commute.

For this we will need two results about symplectic fibrations which are sharpened versions of results proved in Chapter 1: Let $\pi: X \to B$ be a symplectic fibration with compact fibers. (X and B themselves, however, need not be compact.) Let ω_B be a symplectic form on B and let Γ_i, $i = 0, 1$, be symplectic connections on X and ω_{Γ_i} the coupling form associated with Γ_i.

Theorem 4.6.1 *Let U be an open subset of B whose closure is compact. Then there exists a smooth isotopy $f_\epsilon: X \to X$, which is the identity for $\epsilon = 0$, such that, for ϵ sufficiently small,*

$$f_\epsilon^*(\epsilon\omega_{\Gamma_1} + \pi^*\omega_B) = \epsilon\omega_{\Gamma_0} + \pi^*\omega_B \qquad (4.56)$$

on the open set $\pi^{-1}(U)$.

Proof. This is true for the same reason that Theorem 1.6.2 is true. (In fact, if B is compact and $U = B$ it *is* Theorem 1.6.2.) □

The other result we will need is the following: Let A be a compact subset of B and let U be an open neighborhood of it. Let Γ be a symplectic connection on $\pi^{-1}(U)$.

Theorem 4.6.2 *There exists a symplectic connection Γ' on X and a neighborhood U' of A contained in U such that $\Gamma' = \Gamma$ on $\pi^{-1}(U')$.*

Proof. For $U = \emptyset$ this theorem just asserts that there exists a symplectic connection on X. We gave a proof of this in Section 1.2 (see Theorem 1.2.5). We leave for our readers the exercise of showing that Theorem 4.6.2 can be proved by a (trivial) adaptation of this proof. □

We also leave as exercises the G-equvariant versions of these theorems: That is, suppose G acts on X and on B so that the fibration π intertwines these two G-actions. Suppose, moreover, that the various objects mentioned earlier, $A, \Gamma, \Gamma_0, \Gamma_1$, and so forth, are G-equivariant. Then the conclusions of Theorems 4.6.1 and 4.6.2 can be made G-equivariant: That is, the isotopy in Theorem 4.6.1 and the connection Γ' in Theorem 4.6.2 can be made G-invariant.

Let's now come back to the reduction problem: As we mentioned earlier, the modifications we are going to make in symplectic structure are going to take place on X; however, we will first describe the subset of Y over which these modifications are going to take place. Let α be a regular value of the moment mapping $\psi: Y \longrightarrow \mathfrak{g}^*$. Then the action of G on $\psi^{-1}(\alpha)$ is locally free; and, to simplify the discussion a little, we'll assume that the action of G on $\psi^{-1}(\alpha)$ is, in fact, *free*. Suppose now that α is contained in an open polytope Δ_i consisting entirely of regular values of ψ. (See Example 3.7.) Then the action of G is free on the whole open set

$$U = \psi^{-1}(\Delta_i).$$

Therefore since G is compact, the foliation of U associated with the G-orbits is *fibrating*: There exists a principal G-fibration

$$p: U \longrightarrow W \qquad\qquad (4.57)$$

whose fibers are the G-orbits in U. Moreover, since ψ is G-invariant, it factors through p: There exists a smooth fiber mapping $\gamma: W \longrightarrow \Delta_i$ that makes the following diagram commute.

$$
\begin{array}{ccc}
U & & \\
\Big\downarrow{\scriptstyle p} & \searrow{\scriptstyle \psi} & \qquad\qquad (4.58)\\
W & \xrightarrow{\ \gamma\ } & \Delta_i
\end{array}
$$

Moreover, the fiber of γ above a point α in Δ_i is just the set

$$\gamma^{-1}(\alpha) \;=\; \psi^{-1}(\alpha)/G;$$

that is, it is, as an abstract manifold, the reduced space N_α. Therefore, each fiber of the fibration $\phi: W \longrightarrow \Delta_i$ has an intrinsic symplectic structure. However, p is *not* a symplectic fibration in the sense of Chapter 1, since the symplectic structure varies from fiber to fiber. (It varies, of course, in a very simple way: See Example 3.8.)

Now let $\pi: X \longrightarrow Y$ be a symplectic fibration in the sense of Section 1.1. (X itself doesn't yet have to have a symplectic structure, but each of the fibers does.) Suppose that the action of G on Y lifts to an action of G on X which preserves the symplectic structures on the fibers. Let's denote the set $\pi^{-1}(U)$ by X_U. We will define a symplectic fiber bundle X_W over W by defining the fiber of X_W over the point w to be the set of all G-orbits in X that sit over the

G orbit $p^{-1}(w)$. We then have a commutative diagram

$$
\begin{array}{ccc}
X_U & \xrightarrow{\ p\ } & X_W \\
\downarrow{\scriptstyle \pi} & & \downarrow{\scriptstyle \pi} \\
U & \xrightarrow{\ p\ } & W
\end{array}
\tag{4.59}
$$

in which the horizontal arrows are principal G-fibrations.

Let us now show that $\pi\colon X_W \to W$ is a symplectic fibration in the sense of Definition 1.2.6. To begin with it is clear that each fiber is equipped with a symplectic form. Indeed let $w \in W$ be a point and let $x \in U$ be any point in $p^{-1}(w)$. Equip $\pi^{-1}(w)$ with the symplectic form it acquires via the identification $\pi^{-1}(x) \simeq \pi^{-1}(w)$. It follows from the G-invariance of the form ω_{X_U} on X_U that the induced form on $\pi^{-1}(w)$ does not depend on the choice of the point x.

Let us now show that there exists a two-form ω_{X_W} on X_W that satisfies the integrability hypothesis (1.5) and whose restriction to each fiber of π is the symplectic form we have just described.

Suppose $V \subset W$ is an open set such that the fibration $p\colon p^{-1}(V) \to V$ is trivial and let $s\colon V \to W$ be a trivializing section. Then $p\colon \pi^{-1}(p^{-1}(V)) \to \pi^1(V)$ is a diffeomorphism. Its inverse $\tilde{s}\colon \pi^1(V) \to \pi^{-1}(p^{-1}(V))$ is a local section of $p\colon X_U \to X_W$. It is clear that the form

$$
\omega_V = \tilde{s}^* \omega_{X_U}
$$

has the properties required of ω_{X_W} on the open set $\pi^{-1}(V)$.

Now choose a covering of W by open sets $\{W_i \mid i \in I\}$ such that the fibration $p\colon p^{-1}(W_i) \to W_i$ is trivial and let $\{\rho_i \mid i \in I\}$ be a partition of unity subordinate to the covering. Consider the form

$$
\sum (\rho_i \circ \pi)\omega_{W_i},
$$

where the forms ω_{W_i}s are defined analogously to the way the form ω_V is defined. It has the required properties on *all* of X_W, making $\pi\colon X_W \to W$ into a symplectic fiber bundle (see Section 1.2 for more details).

Let Γ^0 be a symplectic connection on the bundle $\pi\colon X_W \longrightarrow W$. By the operation of "base-change" (Example 2.4) it induces a symplectic connection

$$
\Gamma = p^*\Gamma^0
\tag{4.60}
$$

on the bundle $\pi\colon X_U \longrightarrow U$. Let ω_Γ and ω_{Γ^0} be the coupling forms associated with these connections. Then by (4.60)

$$
\omega_\Gamma = p^*\omega_{\Gamma^0},
\tag{4.61}
$$

and so, in particular, if ξ is an element of $\mathfrak{g}$ and $\xi^{\#}$ the vector field it induces on X,

$$\iota(\xi^{\#})\omega_\Gamma = 0. \tag{4.62}$$

Therefore if ω_U is the restriction to U of the symplectic form on Y, the weak coupling form $\epsilon\omega_\Gamma + \pi^*\omega_U$ satisfies

$$\iota(\xi^{\#})(\epsilon\omega_\Gamma + \pi^*\omega_U) = \iota(\xi^{\#})\omega_U = d\langle\psi,\xi\rangle. \tag{4.63}$$

In other words:

Theorem 4.6.3 *For all $\epsilon \neq 0$ the moment mapping associated with the weak coupling form $\epsilon\omega_\Gamma + \pi^*\omega_U$ is $\psi \circ \pi$. (In particular it is independent of ϵ.)*

This achieves the goal we set out to achieve: By changing the symplectic structure on the open subset X_U of X we have forced the diagram (4.54) to commute. Moreover, we *haven't* changed the symplectic structure on X_U very drastically: Replace U by a slightly smaller open set, say V, whose closure in U is compact. Then by Theorems 4.6.1 and 4.6.2 the weak coupling form $\epsilon\omega_\Gamma + \pi^*\omega_U$ associated with the new symplectic structure is, for ϵ sufficiently small, symplectomorphic on V to the weak coupling form associated with the original symplectic structure.

It's now easy to compute the reduced spaces M_α and N_α in (4.54). Consider the diagram:

$$\begin{array}{ccc} X_W & & \\ {\scriptstyle \pi}\downarrow & \searrow {\scriptstyle \gamma\circ\pi} & \\ W & \xrightarrow{\ \gamma\ } & \Delta_i \end{array} \tag{4.64}$$

In this diagram the fiber of γ over α is the reduced space N_α. It is clear from the diagram (4.59) that, at least set-theoretically, the fiber of $\gamma \circ \pi$ over α is the reduced space M_α. Restricting the vertical arrow in (4.64) to $\pi^{-1}(N_\alpha) = M_\alpha$ one gets a fibration

$$\pi_\alpha\colon M_\alpha \longrightarrow N_\alpha. \tag{4.65}$$

Consider, finally, the diagram

$$\begin{array}{ccc} \phi^{-1}(\alpha) & \xrightarrow{\ p\ } & M_\alpha \\ {\scriptstyle i}\downarrow & & \downarrow{\scriptstyle i} \\ X_U & \xrightarrow{\ p\ } & X_W \end{array} \tag{4.66}$$

the is being the inclusion maps. The connection Γ^0 gives rise (by base-change) to a symplectic connection Γ_α on the fiber bundle (4.65); and the minimal coupling forms associated with Γ^0 and Γ_α are related by

$$\omega_{\Gamma_\alpha} = i^*\omega_{\Gamma^0}. \tag{4.67}$$

Let ω_α be the reduced symplectic form on the reduced space N_α. Then, putting together (4.61), (4.66), and (4.67), we get

$$p^*(\epsilon\omega_{\Gamma_\alpha} + \pi_\alpha^*\omega_\alpha) = i^*(\epsilon\omega_\Gamma + \pi^*\omega_U), \tag{4.68}$$

which is exactly the statement *that $\epsilon\omega_{\Gamma_\alpha} + \pi_\alpha^*\omega_\alpha$ is the reduced symplectic form on the reduced space M_α.*

4.7 Some Final Comments about Weak Coupling

A lot of the results that have been discussed in this chapter have been formulated in terms of the weak coupling limit. In this section we will discuss two concrete situations in which weak coupling limits arise. The first has to do with fibrations of coadjoint orbits by coadjoint orbits. (See Section 2.3. The references to follow will be to material in this section.) Let G be a connected Lie group and μ a split element of $\mathfrak{g}^*$. Let's denote by $\mathcal{O}_\mu$ the orbit of G through μ and by G_μ the stabilizer group of μ in G. Then, as a homogeneous space, $\mathcal{O}_\mu$ is just the coset space G/G_μ and the point–coset map

$$\kappa: G \longrightarrow \mathcal{O}_\mu \tag{4.69}$$

makes G into a principal bundle with structure group G_μ. In addition, the left action of G on itself makes G into a group of bundle automorphisms of this bundle. Since μ is split, there exists, by definition, a G_μ-invariant splitting

$$\mathfrak{g} = \mathfrak{g}_\mu \oplus \mathfrak{n}_\mu \tag{4.70}$$

and, therefore, by duality, a G_μ-invariant splitting

$$\mathfrak{g}^* = \mathfrak{g}_\mu^* \oplus \mathfrak{n}_\mu^*,$$

from which we get a G_μ-invariant embedding

$$\mathfrak{g}_\mu^* \longrightarrow \mathfrak{g}^*. \tag{4.71}$$

Therefore, in the subsequent discussion we can think of $\mathfrak{g}_\mu^*$ as sitting inside of $\mathfrak{g}^*$.

The splitting (4.70) also defines a connection on the bundle (4.69) which is invariant under the action of G (as bundle morphisms). Now let f be an element of $\mathfrak{g}_\mu^*$ and $\mathcal{O}$ the orbit of G_μ through f. Starting with the principal bundle (4.69), we can form the associated bundle

$$W = G \times \mathcal{O} / \sim, \tag{4.72}$$

where "$\sim$" is the equivalence relation:

$$(gh, v) \sim (g, hv) \quad \text{for all } h \in G_\mu.$$

It's clear that the map

$$G \times \mathcal{O} \xrightarrow{pr_1} G \xrightarrow{\kappa} \mathcal{O}_\mu$$

factors through "$\sim$", and so gives rise to a fiber mapping

$$\pi: W \longrightarrow \mathcal{O}_\mu \tag{4.73}$$

with typical fiber $\mathcal{O}$. Since $\mathcal{O}$ is a symplectic manifold, this is a symplectic fibration. Moreover, the connection on (4.69) induces a connection on (4.73), say Γ, and thus gives rise to a coupling form ω_Γ. Let $\omega_{\mathcal{O}_\mu}$ be the Kostant–Kirillov symplectic form on $\mathcal{O}_\mu$ and let

$$\omega = \omega_\Gamma + \pi^* \omega_{\mathcal{O}_\mu}. \tag{4.74}$$

The form (4.74) is a symplectic form if and only if the fatness hypothesis of Theorem 2.3.3 is valid. We saw in the course of the proof of Theorem 2.3.7 that this hypothesis is equivalent to the bilinear form

$$\mathfrak{n}_\mu \times \mathfrak{n}_\mu \xrightarrow{\text{Lie bracket}} \mathfrak{g} \xrightarrow{\mu+f} \mathbb{R} \tag{4.75}$$

being nondegenerate. Notice by the way that this condition is automatically satisfied if $f = 0$. Indeed, let $f = 0$ and suppose a vector $\xi \in \mathfrak{n}_\mu$ is annihilated by (4.75). Then, for all $\eta \in \mathfrak{n}_\mu$,

$$0 = \langle \mu, [\xi, \eta] \rangle = -\langle \mu, (\text{ad } \eta)\xi \rangle = -\langle (\text{ad } \eta)^* \mu, \xi \rangle.$$

On the other hand, if $\eta \in \mathfrak{g}_\mu$, then

$$\langle \mu, [\xi, \eta] \rangle = -\langle \mu, (\text{ad } \eta)\xi \rangle = -\langle (\text{ad } \eta)^* \mu, \xi \rangle = 0,$$

so, by (4.70), $\langle \mu, [\xi, \eta] \rangle = 0$ for all $\eta \in \mathfrak{g}$. Hence $\text{ad}(\xi)^* \mu = 0$. Thus ξ is in $\mathfrak{g}_\mu$. Since ξ is in $\mathfrak{n}_\mu$ by assumption and $\mathfrak{g}_\mu \cap \mathfrak{n}_\mu = 0$, $\xi = 0$. This proves that (4.75) is nondegenerate if f is zero. Since (4.75) depends linearly on f it follows that (4.75) is nondegenerate when f is sufficiently small.

Let's come back to the action of G on all the foregoing objects. The action of G on the bundle (4.69) preserves the connection associated with the splitting (4.70). Hence the induced action of G on the bundle (4.73) preserves the connection Γ and the symplectic form (4.74). Moreover, since G acts transitively on the bundle (4.69), and G_μ acts transitively on $\mathcal{O}$, G acts transitively on the manifold W. Thus if the fatness condition (4.75) is satisfied W has to be either a coadjoint orbit of G or a covering of a coadjoint orbit. In fact, since the fiber of the fibration (4.73) is the coadjoint orbit $\mathcal{O}$ of G_μ, W is the coadjoint orbit of G through $\mu + f$ by Theorem 2.3.3.

It is clear from the preceding discussion what happens if we replace the symplectic form (4.74) by the weak coupling form

$$\epsilon \omega_\Gamma + \pi^* \omega_{\mathcal{O}_\mu}. \tag{4.76}$$

The effect of rescaling the symplectic form on $\mathcal{O}$ by ϵ is to replace the orbit $\mathcal{O}$ by the orbit $\epsilon \mathcal{O}$, that is, to replace f by ϵf. Thus if we equip W with the symplectic form (4.76), it becomes the symplectic manifold $\mathcal{O}_{\mu+\epsilon f}$: the coadjoint orbit of G through $\mu + \epsilon f$. Thus, to summarize, we have proved the following:

Theorem 4.7.1 *Let μ and ν be elements of $\mathfrak{g}^*$, μ being semisimple, and let $\mathcal{O}_\mu$ and $\mathcal{O}_\nu$ be the coadjoint orbits containing them. Suppose there exists a G-equivariant map $\pi \colon \mathcal{O}_\nu \longrightarrow \mathcal{O}_\nu$ mapping ν onto μ. Then π has to be a symplectic fibration. Moreover, the symplectic manifold that one obtains by rescaling the symplectic forms on the fibers by ϵ is the coadjoint orbit through $(1 - \epsilon)\mu + \epsilon\nu$.*

For example, let $G = U(n)$. Let $\mathcal{W}_{\lambda_0}$ and $\mathcal{W}_\lambda$ be the coadjoint orbits associated with the sequences $\lambda_0 = (a, \ldots, a, b)$ and $\lambda = (\lambda_1, \ldots, \lambda_n)$. (See (4.17).) Then $\mathcal{W}_{\lambda_0} = \mathbb{C}P^{n-1}$ (the symplectic form on $\mathcal{W}_{\lambda_0}$ being, up to scaling, the standard Fubini–Study symplectic form ω_{FS}). In this example the operation of weak coupling amounts to replacing $\mathcal{W}_\lambda$ by $\mathcal{W}_{\lambda_\epsilon}$, where

$$\lambda_\epsilon = (1 - \epsilon)\lambda_0 + \epsilon\lambda.$$

Next let's turn to the second example of weak coupling that we referred to earlier: Let M be a compact symplectic manifold. Suppose the circle group S^1 acts on M in a Hamiltonian fashion and let

$$\phi \colon M \longrightarrow \mathbb{R}$$

be the moment map associated with this action.

The image of ϕ has to be a compact interval, and we can assume without loss of generality that the bottom endpoint of this interval is the origin. Let $X = \phi^{-1}(0)$. By the Atiyah connectivity theorem, X is a connected submanifold of M, and since $d\phi_x = 0$ at all points $x \in X$, X is a connected component of the fixed-point set of S^1 and, as such, is a symplectic submanifold of M.

If p is a point in M that is near X, but not on X, the action of S^1 at p has to be *locally free*, that is, the stabilizer group of p has to be a finite subgroup of S^1. Henceforth, we will assume that for all such points p the action of S^1 at p is *free*: That is, we will assume that there exists a neighborhood U of X such that for all $p \in U$ either the stabilizer group of S^1 is trivial or is S^1 itself (in which case p has to be on X). Without loss of generality we can assume U is a neighborhood of the form

$$U_\delta = \phi^{-1}([0, \delta))$$

(and, in particular, that it is S^1-invariant).

We will now describe very precisely how S^1 acts on U_δ: Given $p \in X$ let N_p be the normal space to X at p. Then there exists a unique Hermitian form H_p on N_p such that

$$\operatorname{Im} H_p = \omega_p$$

and

$$\operatorname{Re} H_p = d^2\phi_p,$$

$d^2\phi_p$ being the Hessian of ϕ at p and ω_p the restriction of the symplectic form to N_p. (Remember that N_p sits inside the tangent space of M to p as the symplectic orthogonal complement of the tangent space to X at p.) Let Z_p be the set of frames

$$\{(e_1, \ldots, e_n) \mid e_i \in N_p, \ H(e_i, e_j) = \delta_{ij}\}$$

and let

$$Z \longrightarrow X \tag{4.77}$$

be the principal $U(n)$-bundle with fiber Z_p at p. Let's apply, in this setting, the constructions described in Sections 2.2 and 2.3: The standard linear action of $U(n)$ on $\mathbb{C}^n$ preserves the symplectic form $\frac{1}{\sqrt{-1}}\partial\bar\partial|z|^2$; hence we can think of $\mathbb{C}^n$ as being a Hamiltonian $U(n)$-space. We won't bother to describe the moment mapping associated with the action of $U(n)$ on $\mathbb{C}^n$ except to observe that its restriction to the center S^1 of $U(n)$ is the map

$$\phi(z) = |z|^2. \tag{4.78}$$

Given the principal $U(n)$-bundle (4.77) and the action of $U(n)$ on $\mathbb{C}^n$ we can form the associated bundle

$$(Z \times \mathbb{C}^n)/U(n), \tag{4.79}$$

which is, of course, just the normal bundle NX in a new guise. However, there is an advantage to thinking of NX as being the object (4.79): Namely, we can apply the results of Section 2.3 to it.

In particular, if we equip the bundle (4.77) with a connection, the bundle NX acquires a symplectic connection Γ and, hence, a presymplectic two-form

$$\omega_\Gamma + \pi^* \omega_X. \tag{4.80}$$

(For the moment, by the way, we are forgetting about the fact that NX is a vector bundle and are thinking of it as simply being a *manifold*.) The action of the center S^1 of $U(n)$ on $\mathbb{C}^1$ commutes with the action of $U(n)$ itself; so, by (4.79), it gives rise to a *symplectic* action of S^1 on NX. It is easy to see that this is identical with the *linear isotropy action* of S^1 on NX, and, by (4.78), that the moment map associated with this action is the map

$$(p, v) \longrightarrow H_p(v).$$

Now let

$$U_\delta = \{(p, v) \mid H_p(v) \le \delta\}.$$

If δ is small enough then Theorem 2.2.1 tells us *that U_δ is symplectomorphic as an S^1-space to the space we were calling U_δ earlier, namely, the set* $\phi^{-1}([0, \delta))$. The construction we've just outlined can be thought of *as giving us a symplectic canonical form for* $\phi^{-1}([0, \delta))$.

Let's now look at the *reduction problem for U_δ*. For the model space $\mathbb{C}^n$, the reduction operation gives rise to $\mathbb{C}P^{n-1}$s: If we set the right-hand side of (4.78) equal to ϵ and divide out by the circle group orbits

$$\{e^{i\theta} z \mid \quad e^{i\theta} \in S^1\}$$

we obtain $\mathbb{C}P^{n-1}$ equipped with the symplectic form $\epsilon \omega_{FS}$. Therefore, given the description (4.79) of NX, if we apply the same operation to NX, we obtain a presymplectic manifold which, as an abstract space, is just

$$(Z \times \mathbb{C}P^{n-1})/U(n).$$

This space fibers over X and the typical fiber is the symplectic manifold $\mathbb{C}P^{n-1}$. Moreover, the connection Γ on Z gives rise to a symplectic connection on this space and, hence, gives rise to a presymplectic structure of the form (4.80), however, with an ϵ in front of ω_Γ (since the fiber $\mathbb{C}P^{n-1}$ is equipped with the form $\epsilon \omega_{FS}$). In other words, the symplectic form on our space is the *weak coupling* form

$$\epsilon \omega_\Gamma + \pi^* \omega_X. \tag{4.81}$$

Now let ϵ be on the interval $(0, \delta)$. As we've just seen, the open subset

$$\{(p, v) \mid p \in X, \ v \in N_p, \ H_p(v) < \delta\}$$

of NX is a symplectic model for the open subset U_δ of M. Therefore, its reduction with respect to S^1 is, for any ϵ on the interval $(0, \delta)$, symplectomorphic to the corresponding reduced space M_ϵ of M.

Thus to summarize we've proved:

Theorem 4.7.2 *For ϵ on the interval $(0, \delta)$, M_ϵ fibers symplectically over X with fiber $\mathbb{C}P^{n-1}$, and its symplectic form is the weak coupling form (4.81).*

This result has an equivariant generalization: Suppose that H is an N-torus and that H acts in a Hamiltonian fashion on M. Suppose, moreover, that this action commutes with the action of S^1. Then H acts on M_ϵ and on X; and the fibration of M_ϵ over X is a fibration of H-spaces in the sense of (1.1). Thus the results of Section 4.6 can again be of use. To illustrate how, we will outline an inductive procedure for computing the M_αs associated with the Δ_is of Example 3.7. (Incidentally, the notation we'll be using will be that of Section 3.1.) Our procedure will be of limited applicability. It will only apply to those Δ_is that contain in their closures an $(n-1)$-dimensional exterior face of the polytope Δ. Let Δ' be such a face and let η be its inward-pointing normal vector. Since Δ' sits in $\mathfrak{g}^*$, η sits in $\mathfrak{g}$ and is the generator of a circle subgroup S^1 of G. This subgroup acts in a Hamiltonian fashion on M and its moment map is $\langle \Phi, \eta \rangle$. Without loss of generality we can assume that the linear functional $\mathfrak{g}^* \ni \mu \longrightarrow \langle \mu, \eta \rangle$ takes the value zero on Δ', that is,

$$\Delta' = \{\mu \in \Delta \mid \langle \mu, \eta \rangle = 0\}.$$

Let X be the set

$$\{p \in M \mid \phi(p) \in \Delta'\}. \tag{4.82}$$

As we've seen already, X is a connected symplectic submanifold of M. Moreover, confining our attention for the moment to the action of S^1 on M, we get, by the preceding discussion, a reduced space M_ϵ for every ϵ on the interval $(0, \delta)$ and a symplectic fibration of this space over X

$$\pi: M_\epsilon \longrightarrow X$$

with typical fiber $\mathbb{C}P^{N-1}$ ($2N$ being the codimension of X in M).

Let H be a complementary torus to S^1 in G. The action of H on X has, as its moment mapping, the restriction of ϕ to X; and the image of this mapping is Δ' by (4.82). Let Δ'_0 be the set of regular values of the mapping

$$\phi: X \longrightarrow \Delta'$$

and let

$$\Delta'_0 = \bigcup \Delta'_i$$

be the decomposition of Δ'_0 into convex polytopical subsets described in Section 3.1. Let α be in one of these Δ_is. Viewing M_ϵ and X as H-spaces, we can reduce them with respect to α; and by the results of Section 4.6, the reduced spaces that we obtain, say M_α and N_α, are symplectic manifolds, and there exists a symplectic fibration of one over the other

$$\pi: M_\alpha \longrightarrow N_\alpha$$

with typical fiber $\mathbb{C}P^{N-1}$.

Let β be the element of $\mathfrak{g}^*$ defined by the equations:

$$\begin{aligned} \beta(\eta) &= \epsilon, \\ \beta(\xi) &= \alpha(\xi) \qquad \text{for } \xi \in \mathfrak{h}^*, \end{aligned} \tag{4.83}$$

and let M_β be the reduction of M (viewed as a G-space) with respect to β. By "reduction in stages" M_α and M_β are isomorphic as symplectic manifolds, so the previous fibration becomes a symplectic fibration

$$M_\beta \longrightarrow N_\alpha. \tag{4.84}$$

We will conclude this section by describing a couple of special cases of fibrations of this form.

Case 1. Suppose the action of H on X is a *Delzant* action, that is, suppose that

$$2 \dim H = \dim X = 2 \dim \Delta'. \tag{4.85}$$

Then the reduced space N_α is a point and so the corresponding reduced space M_β of G is symplectomorphic to $\mathbb{C}P^{N-1}$. For example, let M be a ten-dimensional orbit of $SU(4)$ and let G be the diagonal subgroup of $SU(4)$. Then the image of M in $\mathfrak{g}^*$ is a truncated tetrahedron with eight exterior faces, four of them hexagons and four of them triangles. (See Section 4.3.) The Xs corresponding to the triangular faces are of Delzant type, and so, for points in regions contiguous to these faces, the corresponding reduced spaces are $\mathbb{C}P^2$s.

Case 2. Let λ be a sequence of type (4.17) and let $\mathcal{W}_\lambda$ be the corresponding coadjoint orbit of $U(n)$. As in Section 4.3 we will think of $\mathcal{W}_\lambda$ as being the set of all $n \times n$ Hermitian matrices with spectrum $\{\lambda_1, \ldots, \lambda_n\}$. Let's assume that

$\mathcal{W}_\lambda$ is a *generic* coadjoint orbit, that is, that the λ_is in the previous sequence are all distinct. As in Section 4.3 let T^n be the diagonal subgroup of $U(n)$; let

$$T^n = T^{n-1} \times S^1, \tag{4.86}$$

where

$$T^{n-1} = \{(e^{i\theta_1}, \ldots, e^{i\theta_{n-1}}, 1)\} \tag{4.87}$$

and

$$S^1 = \{(1, \ldots, 1, e^{i\theta})\}. \tag{4.88}$$

The infinitesimal generator of the group (4.88) is the Hermitian matrix associated with the projection operator

$$P_n: \mathbb{C}^n \longrightarrow \mathbb{C}^n, \qquad v \mapsto (v, e_n)e_n, \tag{4.89}$$

e_n being the vector $(0, \ldots, 0, 1)$; so the moment map associated with the action of (4.88) on $\mathcal{W}_\lambda$ is the map

$$\phi_n: \mathcal{W}_\lambda \longrightarrow \mathbb{R}, \qquad \phi_n(A) = \text{trace } (AP_n).$$

The critical values of this map are $\lambda_1, \ldots, \lambda_n$, and the set of critical points in $\mathcal{W}_\lambda$ associated with the critical value λ_i is the set

$$C_i = \{A \in \mathcal{W}_\lambda \mid \quad Ae_n = \lambda_i e_n\}. \tag{4.90}$$

In particular, one can identify the critical set where ϕ_n takes its *minimum* value with the coadjoint orbit $\mathcal{W}_{\lambda'}$ of $U(n - 1)$, where $\lambda' = (\lambda_1, \ldots, \lambda_{n-1})$. (See Section 4.3.)

Let's now look at the normal bundle to this critical manifold in $\mathcal{W}_\lambda$. The matrix

$$A_\lambda = \begin{pmatrix} \lambda_1 & 0 & \cdots & 0 \\ 0 & \lambda_2 & \cdots & 0 \\ \vdots & \vdots & & \vdots \\ 0 & 0 & \cdots & \lambda_n \end{pmatrix}$$

is the unique point on this critical set stabilized by T^n, and it is easy to see that the normal space to the critical manifold at this point is $\mathbb{C}^{n-1}$. The isotropy action of T^n on this space is also easy to describe in terms of the factorization (4.88). For T^{n-1} it is the *standard* action of T^{n-1} on $\mathbb{C}^{n-1}$ (T^{n-1} being embedded in $U(n - 1)$ as the diagonal subgroup). For S^1, on the other hand, it is the action $e^{i\theta} \longrightarrow e^{-i\theta}$ times the identity. (The latter is easy to see from the fact that the *center* of $U(n)$ has to act trivially.)

As we noted before, this critical manifold can be identified with the $U(n-1)$ orbit $\mathcal{W}_{\lambda'}$. In particular, $U(n-1)$ acts *transitively* on this set, so the normal bundle to this set is just the homogeneous vector bundle

$$E_\rho \longrightarrow \mathcal{W}_\lambda = U(n-1)/T^{n-1} \tag{4.91}$$

induced by the representation ρ, of T^{n-1} on $\mathbb{C}^{n-1}$ that we've just described. Note, however, that this representation extends to a representation of $U(n-1)$ on $\mathbb{C}^{n-1}$. This implies that (4.91) is *trivial* as a $U(n-1)$-bundle, that is, is just the product of $U(n-1)$-spaces

$$\mathcal{W}_{\lambda'} \times \mathbb{C}P^{n-2},$$

$\mathcal{W}_{\lambda'}$ being equipped with its canonical symplectic structure and $\mathbb{C}P^{n-2}$ with the symplectic structure associated with $\epsilon\omega_{FS}$. Thus, by our theorem on induction in stages, we obtain the result:

Theorem 4.7.3 *If* $\alpha = (\alpha_1, \ldots, \alpha_n)$ *and* $\alpha_1 \approx \lambda_n$, *the reduced space* M_α *is the product:*

$$M_{\alpha'} \times (\mathbb{C}P^{n-2})_{\epsilon_{n-2}}, \tag{4.92}$$

where $M_{\alpha'}$ *is the reduction of* $\mathcal{W}_{\lambda'}$ *with respect to* $\alpha' = (\alpha_1, \ldots, \alpha_{n-1})$ *and* $(\mathbb{C}P^{n-2})_\epsilon$ *is the projective space* $\mathbb{C}P^{n-2}$ *equipped with the symplectic form* $\epsilon\omega_{FS}$. *In (4.92),* $\epsilon_{n-2} = \alpha_n - \lambda_n$.

By iterating this result one arrives at the following theorem:

Theorem 4.7.4 *Let* λ *be a sequence of the form (4.17) and let* Δ *be the convex hull of the Weyl group orbit of* λ, *in other words, the convex hull in* $\mathbb{R}^n$ *of the* $n!$ *vectors*

$$(\lambda_{\sigma(1)}, \ldots, \lambda_{\sigma(n)}), \qquad \sigma \in \Sigma_n.$$

Let $\alpha = (\alpha_1, \ldots, \alpha_n)$ *be an element in the convex hull with* $\alpha_3 > \lambda_3, \ldots, \alpha_n > \lambda_n$. *If the differences* $\epsilon_1 = \alpha_3 - \lambda_3, \ldots, \epsilon_{n-2} = \alpha_n - \lambda_n$ *are sufficiently small the reduced space* M_α *is the product:*

$$(\mathbb{C}P^1)_{\epsilon_1} \times \cdots \times (\mathbb{C}P^{n-2})_{\epsilon_{n-2}}. \tag{4.93}$$

Recall, by the way, that we have already shown that if λ is a sequence of the form (4.17) that is "ϵ-close" to the sequence $(a, \ldots, a, b)$, $a > b$, and if Δ_i is the region (in the subdivision of Δ described in Section 3.1) that contains 0, then, for all $\alpha \in \Delta_i$, M_α is symplectomorphic to $\mathcal{W}_{\lambda'}$. As we pointed out in Section 4.1, there is a natural notion of *birational equivalence* for symplectic manifolds; and, in particular, for αs in different Δ_is the corresponding M_αs are birationally equivalent as symplectic manifolds. In particular, it follows from the preceding computations that the "general flag variety" $\mathcal{W}_\lambda$ and the product

$$(\mathbb{C}P^1)_{\epsilon_1} \times \cdots \times (\mathbb{C}P^{n-1})_{\epsilon_{n-1}}$$

are birationally equivalent as symplectic manifolds! (The fact that the general flag variety $\mathcal{W}_n$ and the product $\mathbb{C}P^1 \times \cdots \times \mathbb{C}P^{n-1}$ are birationally equivalent as *complex* manifolds is well known and is easy to prove: Consider the sequence of subspaces

$$U_i, \qquad i = 1, \ldots, n-1,$$

of $\mathbb{C}^n$ defined by the equations

$$z_1 = \cdots = z_{n-1} = 0.$$

Let $(V_1, V_2, \ldots, V_{n-1})$ be a flag with the property that

$$\dim (V_i \cap U_i) = 1$$

for all i. The set of all such flags is a Zariski open subset $\mathcal{O}$ of $\mathcal{W}_n$. If we identify $\mathbb{C}P^k$ with the set of all one-dimensional subspaces of U_{n-k}, we have a bijective map of $\mathcal{O}$ onto a Zariski open subspace $\mathcal{O}'$ of the product

$$\mathbb{C}P^1 \times \cdots \times \mathbb{C}P^{n-1}.$$

Namely, it is the map that sends the foregoing flag onto the $(n-1)$-tuple

$$(U_{n-1} \cap V_{n-1}, \ldots, U_1 \cap V_1).$$

It is easy to see that this map is biholomorphic and that the complements of $\mathcal{O}$ in $\mathcal{W}_n$ and of $\mathcal{O}'$ in $\mathbb{C}P^1 \times \cdots \times \mathbb{C}P^{n-2}$ are subvarieties of (complex) codimension ≥ 2. Thus, this map is a birational equivalence.)

5

Computations with Orbits

5.1 Generalities about Toral Moment Maps

Recall the description of singular points of a moment map in terms of the group action. Let (M, ω) be a compact Hamiltonian space for a compact Lie group G with moment map $\Phi: M \to \mathfrak{g}^*$. Infinitesimally the definition of the moment map says that for any ξ in $\mathfrak{g}$, the Lie algebra of G, and any vector v tangent to M

$$\langle d\Phi_m(v), \xi \rangle = \omega_m(\xi_M(m), v). \tag{5.1}$$

Here $\langle \, , \, \rangle$ denotes the pairing between the Lie algebra $\mathfrak{g}$ and its dual $\mathfrak{g}^*$, and ξ_M is the vector field on M induced by the action of ξ. Equation (5.1) implies that the annihilator $(d\Phi(T_m M))^\circ$ of the image of $d\Phi_m$ in $\mathfrak{g}^*$ satisfies

$$(\operatorname{Im} d\Phi_m)^\circ = \{\xi \in \mathfrak{g} \mid \xi_M(m) = 0\}. \tag{5.2}$$

The right-hand side of (5.2) is the Lie algebra of the stabilizer group G_m of m. Consequently, $d\Phi_m$ is surjective if and only if the stabilizer of m is discrete. Alternatively,

$$m \text{ is a singular point of } \Phi \iff \dim G_m \geq 1. \tag{5.3}$$

To avoid considering degenerate cases let us now assume that the action of G is locally free at some point of M (hence is locally free generically). In general, for any subgroup H of G, its fixed-point set

$$\operatorname{Fix}(H) = \{m \in M \mid a \cdot m = m \text{ for any } a \in H\}$$

is a union of symplectic submanifolds (perhaps of different dimensions); see [GS2]. Therefore the critical points of Φ are a union of symplectic submanifolds of M of the form $\operatorname{Fix}(H)$, where H is a subgroup of G, and the singular values of Φ are the images of these submanifolds.

Assume from now on that G is a torus. Then the fixed points of the action characterize the image completely: $\Phi(M)$ is the convex hull of the images of fixed points. This is the content of the convexity theorem of Atiyah, Guillemin, and Sternberg. It follows easily from the theorem and the preceding discussion that the set of singular values of Φ is a union of (not necessarily disjoint) convex codimension-one polytopes. We will call them the *singular polytopes* of Φ. The singular polytopes are images of fixed-point sets of circle subgroups of G.

In the next section we will see that these polytopes have a particularly nice description when M is a coadjoint orbit $\mathcal{O}$ of a compact Lie group K and G is a maximal torus T of the group K.

5.2 Singular Values of the Moment Map $\Phi\colon \mathcal{O} \to \mathfrak{t}$

In this section we compute the singular points and singular values of the moment map arising from the action of a maximal torus T on a coadjoint orbit $\mathcal{O}$ of a compact Lie group K. The main result of the section, Theorem 5.2.1, is not new. For example, it has been proved by Heckman in his thesis [Hec]. Unlike Heckman, who used Morse theory, we use elementary symplectic geometry to deduce the result.

We start out by recalling a few useful facts about the structure of compact Lie groups. Let K be a compact, connected semisimple Lie group, T its maximal torus, and W the Weyl group of the pair (K, T). The Lie algebra $\mathfrak{k}$ of K carries an invariant positive-definite inner product $(\cdot, \cdot)$, which allows us to identify the Lie algebra $\mathfrak{k}$ with its dual $\mathfrak{k}^*$. This identification intertwines the adjoint and coadjoint actions of K and thereby sets up a correspondence between adjoint and coadjoint orbits. From now on we shall suppress the distinction between $\mathfrak{k}$ and $\mathfrak{k}^*$. Fix a system of positive roots Δ^+. The root space decomposition of $\mathfrak{k}$ is orthogonal with respect to the inner product $(\cdot, \cdot)$ and can be written as follows ([Hel], ch. IV, Lemma 3.1):

$$\mathfrak{k} = \mathfrak{t} \oplus \sum_{\alpha \in \Delta^+} Z_\alpha \tag{5.4}$$

(the compact form of $\mathfrak{k}_{\mathbb{C}}$). Here the subspaces $Z_\alpha = \mathbb{R}u_\alpha \oplus \mathbb{R}v_\alpha$ are two-dimensional and for all x in $\mathfrak{t}$ the basis vectors $\{u_\alpha, v_\alpha\}$ satisfy the equations

$$\begin{aligned}
[x, u_\alpha] &= \alpha(x)v_\alpha, \\
[x, v_\alpha] &= -\alpha(x)u_\alpha, \\
[u_\alpha, v_\alpha] &= (u_\alpha, u_\alpha)\alpha = (v_\alpha, v_\alpha)\alpha.
\end{aligned} \tag{5.5}$$

In the last line α, instead of being a linear functional, denotes, by abuse of notation, a vector in $\mathfrak{t}$ defined by $(\alpha, y) = \alpha(y)$ for all y in $\mathfrak{t}$.

It is easy to see that for x in $\mathfrak{t}$ the kernel of ad $x \colon \mathfrak{k} \to \mathfrak{k}$ satisfies

$$\ker(\operatorname{ad} x) = \mathfrak{t} \oplus \sum_{\alpha \in \Delta^+,\, \alpha(x)=0} Z_\alpha. \qquad (5.6)$$

But $\ker(\operatorname{ad} x) = \{y \in \mathfrak{k} \mid [y, x] = 0\} = \operatorname{Lie}(\operatorname{Stab}(x))$, the Lie algebra of the stabilizer group of x. It follows that T is also a maximal torus of $\operatorname{Stab}(x)$. The roots of $\operatorname{Stab}(x)$ are all the roots α of K with $\alpha(x) = 0$. The Weyl group of $\operatorname{Stab}(x)$ is the subgroup of W generated by reflections in such roots. The smallest Weyl chamber wall containing x is the linear space $\bigcap_{\alpha(x)=0} \ker(\alpha)$. Define the interior of the wall to be the collection of points y that do not lie on any hyperplane $\ker(\beta)$ with $\beta(x) \neq 0$. Then the points in the interior of the wall defined by x have the stabilizers equal to $\operatorname{Stab}(x)$.

This fact is best understood by considering $\mathfrak{t}$ together with all the walls of Weyl chambers as a poset under the reversed inclusion. The smallest element of the poset is $\mathfrak{t}$ itself. The largest element is the origin. The level just below the origin consists of the one-dimensional walls. Each of these lines is defined by a fundamental weight or its Weyl group conjugate. The poset is isomorphic to the poset of the stabilizer groups of points in $\mathfrak{t}$. The correspondence sends a wall of $\mathfrak{t}$ to the stabilizer of an interior point of the wall (the interior of $\mathfrak{t}$ is the union of open Weyl chambers). The Lie algebras of the stabilizer groups and the Weyl groups form isomorphic posets as well.

Given a linear subspace V of $\mathfrak{t}$, its centralizer

$$c(V) = \{w \in \mathfrak{k} \mid [w, x] = 0 \text{ for any } x \text{ in } V\}$$

is a Lie algebra in our poset. It corresponds to the smallest wall containing V. If μ is conjugate to a fundamental weight its stabilizer H_μ is not semisimple: μ spans the one-dimensional center of $\mathfrak{h}_\mu$. The corresponding circle $S_\mu = \{\exp t\mu\}_{t\in\mathbb{R}}$ will play an important role in describing the singular points of the moment map. We finally note that the Dynkin diagram of H_μ is obtained by deleting one vertex from the Dynkin diagram of K.

Let $\mathcal{O} = \mathcal{O}_\lambda$ be the adjoint orbit through a point $\lambda \in \mathfrak{t}$. The identification of $\mathfrak{t}$ with $\mathfrak{t}^*$ makes $\mathcal{O}$ into a Hamiltonian T-space with the Kirillov–Kostant–Souriau symplectic structure. So the group G of Section 5.1 is now called T. Under the identifications the moment map $\Phi \colon \mathcal{O} \to \mathfrak{t}$ is simply the orthogonal projection. The orbit $\mathcal{O}$ intersects $\mathfrak{t}$ in a single Weyl group orbit

$$\mathcal{O} \cap \mathfrak{t} = W \cdot \lambda$$

for some $\lambda \in \mathfrak{t}$. All the points in the intersection of $\mathcal{O}$ and $\mathfrak{t}$ are fixed by the action of the maximal torus T.

Conversely, suppose $y \in \mathcal{O}$ is a fixed point of T. Then for any $x \in \mathfrak{t}$

$$\mathrm{ad}(x)\, y = 0.$$

Consequently y lies in the centralizer of $\mathfrak{t}$ in $\mathfrak{k}$. But the centralizer of $\mathfrak{t}$ is $\mathfrak{t}$ itself, since T is a maximal torus. Therefore y lies in the intersection of $\mathcal{O}$ and $\mathfrak{t}$, and so $y = w \cdot \lambda$ for some element w of the Weyl group W. It is easy to see that no point of the intersection $\mathcal{O} \cap \mathfrak{t}$ lies in the interior of the convex polytope

$$\mathrm{conv}\,(W \cdot \lambda) := \text{ convex hull of } W \cdot \lambda.$$

We conclude that the image of an orbit through λ is the convex hull of its Weyl group orbit

$$\Phi(K \cdot \lambda) = \mathrm{conv}\,(W \cdot \lambda). \tag{5.7}$$

This result was proved by Schur and Horn in the case of $K = SU(n)$ and was later generalized by Kostant for an arbitrary compact group [K3]. In fact equation (5.7) is a special case of Kostant's convexity theorem.

To describe the singular values of the moment map Φ we need to introduce more notation. Let $\mu_1, \ldots, \mu_N$ denote all the fundamental weights and their conjugates, and let $H_i = H_{\mu_i}$ be their stabilizer groups for the (co)adjoint action of K. Denote the Weyl group of the pair (H_i, T) by W_i. The group W_i is generated by reflection in the roots orthogonal to μ_i. Recall that the weight μ_i generates a central circle subgroup $S_i = \{\exp(t\mu_i)\}_{t \in \mathbb{R}}$ of H_i. We saw in (5.3) that any point in $\mathcal{O} = \mathcal{O}_\lambda$ fixed by S_i is a singular point of the moment map Φ. We claim that

1. the fixed-point set of S_i is exactly $\mathcal{O}_\lambda \cap \mathfrak{h}_i$, where $\mathfrak{h}_i = \mathrm{Lie}(H_i)$, and
2. $\Phi(\mathcal{O}_\lambda \cap \mathfrak{h}_i) = \bigcup_{w \in W} \mathrm{conv}\,(W_i \cdot w\lambda)$.

Proof. A point z in $\mathcal{O}_\lambda$ is fixed by S_i if and only if $[\mu_i, z] = 0$. But the Lie algebra $\mathfrak{h}_i$ is exactly the set of all vectors in $\mathfrak{k}$ that commute with μ_i. Therefore (1) holds.

To prove (2) note first that $\mathcal{O}_\lambda \cap \mathfrak{h}_i$ is a union of H_i-orbits passing through the points of the set $W \cdot \lambda$, since every point in $\mathfrak{h}_i$ is $\mathrm{Ad}(H_i)$-conjugate to a point in $\mathfrak{t}$ (recall that T is also a maximal point of H_i). Consider the action of T on the orbit $H_i \cdot w\lambda$ for some $w \in W$. Since $H_i \cdot w\lambda$ is a symplectic submanifold of $\mathcal{O}_\lambda$ the moment map for this action is simply the restriction of Φ to $H_i \cdot w\lambda$. But then by (5.7)

$$\mathrm{image}(\Phi|_{H_i \cdot w\lambda}) = \Phi(H_i \cdot w\lambda) = \mathrm{conv}\,(W_i \cdot w\lambda).$$

Part (2) of the claim now follows. $\square$

Conversely, suppose z is a singular point of Φ. Then it is fixed by some subtorus T' of T (dim $T' \geq 1$). So for any $x \in \mathfrak{t}' = \mathrm{Lie}(T')$ we have

$$[z, x] = 0.$$

That is, z lies in $c(\mathfrak{t}')$, the centralizer of $\mathfrak{t}'$, which in turn is contained in $\mathfrak{h}_i$ for some i. We conclude that $z \in \mathfrak{h}_i$. This proves the following theorem.

Theorem 5.2.1 *The singular points of the moment map $\Phi\colon \mathcal{O}_\lambda \to \mathfrak{t}$ are the symplectic submanifolds*

$$H_i \cdot w\lambda, \qquad w \in W, \ i = 1, \ldots, N.$$

The singular values of Φ are the convex polytopes

$$conv\,(W_i \cdot w\lambda), \qquad w \in W, \ i = 1, \ldots, N.$$

To make the discussion above more concrete let us review the structure of the special unitary group $SU(n)$. Recall that we may identify the Lie algebra $\mathfrak{su}(n)$ of $SU(n)$ with the space of traceless Hermitian $n \times n$ matrices. The Killing form is (a multiple of) the trace form $(A, B) \mapsto \mathrm{tr}\,AB$. The collection of diagonal matrices

$$\begin{pmatrix} e^{i\theta_1} & & \\ & \ddots & \\ & & e^{i\theta_n} \end{pmatrix}$$

with $\prod e^{i\theta_j} = 1$ is a maximal torus T of $SU(n)$. We identify its Lie algebra $\mathfrak{t}$ with the space of traceless diagonal matrices

$$\mathfrak{t} = \left\{ \begin{pmatrix} \theta_1 & & \\ & \ddots & \\ & & \theta_n \end{pmatrix} \;\middle|\; \sum \theta_i = 0 \right\}.$$

We may also identify $\mathfrak{t}$ with a hyperplane in $\mathbb{R}^n$:

$$\mathfrak{t} = \left\{ x \in \mathbb{R}^n \;\middle|\; \sum x_j = 0 \right\}.$$

When considering the root space decomposition of $\mathfrak{su}(n)$, it is convenient to think of the Lie algebra as the space of *skew*-Hermitian matrices. For $1 \leq i < j \leq n$ let u_{ij} be the skew-Hermitian matrix with 1 in the (i, j) entry, -1 in the (j, i) entry, and zeros everywhere else, and let v_{ij} be the matrix with $\sqrt{-1}$ in the (i, j) and (j, i) entries and zeros elsewhere. Then $(u_{ij}, u_{ij}) = -\,\mathrm{tr}\,u_{ij}^2 = 2$

and $(v_{ij}, v_{ij}) = -\operatorname{tr} v_{ij}^2 = 2$. Let Z_{ij} be the $\mathbb{R}$-linear span of $\{u_{ij}, v_{ij}\}$. Then we have

$$[u_{ij}, v_{ij}] = 2e_{ij} = (u_{ij}, u_{ij})e_{ij} = (v_{ij}, v_{ij})e_{ij},$$

where $e_{ij} \in \mathfrak{t}$ corresponds to the n-tuple $(0, \ldots, \overset{i}{1}, \ldots, \overset{j}{-1}, \ldots, 0)$. Also, for any x in $\mathfrak{t}$,

$$[x, u_{ij}] = (x_i - x_j) \cdot v_{ij} = (e_{ij}, x) \cdot v_{ij},$$
$$[x, v_{ij}] = -(x_i - x_j) \cdot u_{ij} = -(e_{ij}, x) \cdot u_{ij}.$$

It follows that the Z_{rs}s are the root spaces of $\mathfrak{su}(n)$. They are indexed by pairs of integers (r, s), $1 \le r < s \le n$; the (r, s) root space Z_{rs} has complex dimension one and consists of the (skew-)Hermitian matrices of the form

$$
\begin{array}{c}
\quad\quad r \quad\quad\quad s \\
\begin{array}{c} r \\ \\ s \end{array}
\left(
\begin{array}{ccc}
0 & \cdots & w \\
\vdots & & \vdots \\
\bar{w} & \cdots & 0
\end{array}
\right),
\end{array}
$$

with a complex number w in the (r, s) entry, $\bar{w}$ in the (s, r) entry, and zeros elsewhere. The corresponding positive root e_{rs} is given by

$$e_{rs}(x_1, \ldots, x_n) = x_r - x_s.$$

If we identify $\mathfrak{t}$ with $\mathfrak{t}^*$ using the restriction of the trace form (under the identification of $\mathfrak{t}$ with a subspace of $\mathbb{R}^n$ this is just the standard inner product), then

$$e_{rs} = (0, \ldots, \overset{r}{1}, \ldots, \overset{s}{-1}, \ldots, 0), \quad 1 \le r < n, \, 1 < s \le n, r < s,$$

are the positive roots. The negative roots are the vectors of the form

$$(0, \ldots, -1, \ldots, 1, \ldots, 0).$$

The splitting

$$\mathfrak{su}(n) = \mathfrak{t} \oplus \sum_{1 \le r < s \le n} Z_{rs} \tag{5.8}$$

is the root space decomposition of $SU(n)$.

With this choice of splitting of roots into positive and negative ones, the simple roots are the vectors $e_{12}, e_{23}, \ldots, e_{n-1,n}$, and the positive Weyl chamber $\mathfrak{t}_+$ is given by

$$\mathfrak{t}_+ = \left\{ (x_1, \ldots, x_n) \in \mathbb{R}^n \,\middle|\, x_1 > x_2 > \cdots > x_n, \quad \sum x_i = 0 \right\}.$$

Its closure $\bar{\mathfrak{t}}_+$ is the set

$$\bar{\mathfrak{t}}_+ = \left\{ (x_1, \ldots, x_n) \in \mathbb{R}^n \,\middle|\, x_1 \geq x_2 \geq \cdots \geq x_n, \quad \sum x_i = 0 \right\}.$$

The reflection defined by the root e_{rs} interchanges the rth and the sth coordinates. It follows that the Weyl group of $SU(n)$ is Σ_n, the symmetric group on n letters, and that it acts on $\mathfrak{t}$ by permuting the coordinates. Finally we note that the fundamental weights $\mu_1, \ldots, \mu_n$ are the linear functionals

$$\mu_k(x_1, \ldots, x_n) = x_1 + \cdots + x_k.$$

Alternatively, $n \cdot \mu_k$ is the vector

$$(\overbrace{l, \ldots, l}^{k}, \overbrace{-k, \ldots, -k}^{l}), \qquad l + k = n.$$

Such vectors span the one-dimensional walls of the Weyl chamber $\mathfrak{t}_+$.

5.3 Examples: $SU(4)$ Orbits

In this section we describe the polytopes that occur as images of orbits of $SU(4)$, the singular polytopes, and the components of regular values of the moment maps.

The torus $\mathfrak{t}$ in $\mathfrak{su}(4)$ has an orthonormal basis

$$\begin{aligned}
v_1 &= (1/2, -1/2, 1/2, -1/2), \\
v_2 &= (1/2, -1/2, -1/2, 1/2), \\
v_3 &= (1/2, 1/2, -1/2, -1/2).
\end{aligned} \tag{5.9}$$

This identifies $\mathfrak{t}$ isometrically with $\mathbb{R}^3$. The positive Weyl chamber C_0 is determined by the system of inequalities:

$$\begin{cases} x_3 > x_1, \\ x_1 > x_2, \\ x_1 > -x_2. \end{cases}$$

With respect to the preceding identification the fundamental weights are as follows: $\mu_1 = (1/2, 1/2, 1/2)$, $\mu_2 = (0, 0, 1/2)$, and $\mu_3 = (1/2, -1/2, 1/2)$. The Weyl chambers of $\mathfrak{su}(4)$ can be pictured by drawing their intersection with the surface of a cube centered at the origin with the sides parallel to the coordinate planes. Take a face of the cube and divide it into four triangles by the diagonal lines. Each triangle is the intersection of a Weyl chamber with the surface of the cube; there are 24 in all. Note that the cube itself does not occur as a convex hull of a Weyl group orbit. Rather its appearance is due to the isomorphism of the Lie algebras $\mathfrak{su}(4)$ and $\mathfrak{o}(6)$.

Remark. In general, the Weyl chambers of $SU(n)$ arise from the barycentric subdivision of the $(n-1)$-simplex. This is because Σ_n is the group of symmetries of the simplex. More concretely it can be seen as follows.

We first identify the maximal torus $\mathfrak{t}$ of $\mathfrak{su}(n)$ with the affine hyperplane in $\mathbb{R}^n$ passing through the points $(1, \ldots, 0), (0, 1, \ldots, 0), \ldots, (0, \ldots, 1)$. These points are the Σ_n-orbit of $(1, 0, \ldots, 0)$ and they define an $(n-1)$-simplex Δ with the center at $\frac{1}{n}(1, \ldots, 1)$. The point $\frac{1}{2}(1, 1, 0, \ldots, 0)$ is the midpoint of the edge joining $(1, 0, \ldots, 0)$ with $(0, 1, \ldots, 0)$; the point $\frac{1}{3}(1, 1, 1, 0 \ldots, 0)$ is the center of the 2-simplex spanned by the vectors $(1, 0, \ldots, 0)$, $(0, 1, \ldots, 0)$, and $(0, 0, 1, \ldots, 0)$, and so on. In other words the fundamental weights pass through the barycenters of the simplex Δ. And conversely, given an $(n-1)$-simplex of the barycentric subdivision of Δ, it spans a cone with the vertex at $\frac{1}{n}(1, \ldots, 1)$. This cone is a Weyl chamber of $SU(n)$.

Let us now discuss the images of orbits of $SU(4)$. The most singular orbits, of course, are the ones passing through (the multiples of) the fundamental weights. The orbits through $\mu_1 = \frac{1}{4}(3, -1, -1, -1) \sim \frac{1}{2}(1, 1, 1)$ and $\mu_2 = \frac{1}{4}(1, 1, 1, -3) \sim \frac{1}{2}(1, -1, 1)$ are projective spaces $\mathbb{C}P^3$. The orbit through $\mu_2 = \frac{1}{4}(1, 1, -1, -1) \sim (0, 0, 1/2)$ is a Grassmannian of two-planes in $\mathbb{C}^4$. It has real dimension 8. The image of the orbit is an octahedron. The singular values of the moment map (besides the boundary points) are the squares formed by the intersections of the polyhedron with the three coordinate planes. The octahedron breaks up into eight little triangular pyramids which have the faces of the octahedron as their bases and share a common vertex at the origin (see Figure 5.1).

Fig. 5.1. Partition of the octahedron.

The singular submanifolds of the Grassmannian corresponding to the faces are $\mathbb{C}P^2$s which are of codimension 4. A weak-coupling argument (cf. Section 4.7) shows that all the reduced spaces are $\mathbb{C}P^1$s. The corresponding D–H measure is linear inside each pyramid. Its level sets are parallel to the base and the measure is increasing toward the origin.

There are two types of 10-dimensional orbits depending on whether the point $x = (x_1, x_2, x_3)$ sits on the wall spanned by the weights μ_1 and μ_3 (i.e., satisfies $x_3 = x_1$ and $x_1 > |x_2|$) – this corresponds to the second and the third eigenvalues being equal – or on one of the other two 2-dimensional walls ($x_1 = \pm x_2$, etc.) – in which case the last two eigenvalues are equal. In the latter case the image of the orbit through x is the truncated tetrahedron mentioned in Chapter 4 (see Figure 5.2).

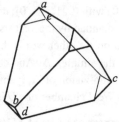

Fig. 5.2. Truncated tetrahedron.

The singular polytopes are either triangles, like the one formed by connecting the points a, b, and c (one for each hexagonal face, four triangles altogether), or rectangles like the one formed by connecting the points a, b, d, and e (one for each long edge, six in all). The set of regular values breaks up into 49 connected components.

The polytope depicted here corresponds to the case of x being close to the line through μ_3. As one moves the point x inside the wall $x_1 = x_2$ away from this line and closer to the line through μ_2 the shape of the image and its decomposition into regions of regular values changes. For example, at some point the hexagonal faces become regular hexagons and then some of the 49 regions collapse. In the next section we will say more about this phenomenon.

Fig. 5.3. Cross section 1.

To get a feel for what these connected components are and how they fit together, we will describe a series of cross sections of the polyhedron by affine planes parallel to the plane abc.

If we cut close to the small triangular face, the cross section looks like Figure 5.3. As the plane moves toward the center, the middle triangle shrinks down

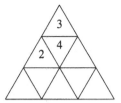

Fig. 5.4. Cross section 2.

to a point (Figure 5.4). This indicates that the regions contingent to the four small triangular faces are triangular pyramids pointing inward. Below this level a triangular central region appears once again in Figure 5.5. It expands until

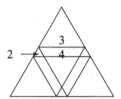

Fig. 5.5. Cross section 3.

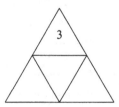

Fig. 5.6. Cross section 4.

it touches the perimeter of the cross section (Figure 5.6) and then begins to shrink (Figure 5.7) all the way down to a point (Figure 5.8). Therefore, below the little pyramid there is a region that is shaped like two triangular pyramids glued along their bases. As we keep moving the plane down toward the singular plane *abc*, a triangle appears in the center once again in Figure 5.9. (Because of the symmetry considerations we do not need to move the cross section past the plane *abc*.) The central triangle increases monotonically. We conclude that the center of the truncated tetrahedron is occupied by a tetrahedron-shaped component.

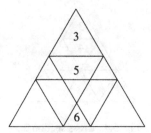

Fig. 5.7. Cross section 5.

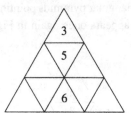

Fig. 5.8. Cross section 6.

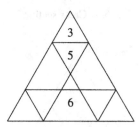

Fig. 5.9. Cross section 7.

Since the image of the orbit has Σ_4 symmetry, we have accounted so far for 9 regions. What about the other 40 regions? As we follow the evolution of the polygon marked "2" on the cross sections 1, 2, and 3 (Figures 5.3–5.5), we see that we are once again slicing a triangular pyramid. There are three such pyramids around each vertex, 12 altogether. One of the faces of such a pyramid lies on the hexagonal face of the truncated tetrahedron.

Similarly, the polygons marked "4" are cross sections of a triangular pyramid. Again, there are three of these polytopes around each vertex, which accounts for another 12 regions in the partition of the polytope. There are 16 regions left to describe.

The triangles marked "3" are cross sections of a slightly more complicated polytope. Note that the triangles increase as the cross section moves to the level of cross section 4 and then remain constant in size. To visualize this polytope start with a triangular prism. Slice off a piece of the prism by a plane through one of the vertices. Keep this plane parallel to an edge of the base about to be sliced off. The result is a lopsided prism with one of the sides still being a rectangle. Perform a similar operation at the other end of the prism. If we put the resulting polytope down flat on its unique rectangular face, the polytope looks like an A-frame tent overhanging slightly at both ends. There are six such lopsided prisms, one for each long edge of truncated tetrahedron. The longest edge of the prism coincides with a long edge of the image.

The polygons marked "5" are cross sections of a similarly shaped lopsided triangular prism. The difference is that the two triangular faces are sloping inward rather than overhanging. There are six such regions.

Finally, the polygons marked "6" are cross sections of a truncated triangular pyramid. The base of this pyramid lies on a hexagonal face of the truncated tetrahedron and the face opposite to the base is contiguous to the central tetrahedral region. In fact, if we glue the two regions together, we will get a full pyramid. There are four such polyhedrons, one for each hexagonal face, which brings the total number of regions to 49.

Of the 49 regions described, 38 have faces lying on the boundary of the polytope: seven regions for each hexagonal face, one region for each triangular face, and one region for each rectangular face. Weak-coupling arguments for these regions (cf. Section 4.7) imply that the reduced spaces are either $\mathbb{C}P^2$s or $\mathbb{C}P^1$-bundles over $\mathbb{C}P^1$s. Applying the fibration of orbits and weak-coupling argument to the central region we conclude that the reduced spaces corresponding to the points of the region are $\mathbb{C}P^2$s and that the D–H measure is constant.

To compute the reduced spaces in the remaining 10 regions one can, for example, use the equivariant Morse-theoretic techniques of [GS9]. The methods imply that the reduced spaces are $\mathbb{C}P^2$s and $\mathbb{C}P^2$s with one, two, or three points blown up. We will explicitly compute the corresponding D–H polynomials in the next section.

Another type of a 10-dimensional orbit corresponds to points in the interior of the wall spanned by the weights μ_1 and μ_3. In terms of the spectrum $\lambda = (\lambda_1, \ldots, \lambda_4)$ this translates into $\lambda_2 = \lambda_3$. If a point x lies exactly halfway between the lines through the two weights then the image of the orbit through x is a cuboctahedron. In general the image is a sort of "skew-cuboctahedron" (Figure 5.10). The singular polytopes are hexagons such as the hexagon $abcdef$. There are four of them. The hexagons break up the polytope into 15 regions, one for each face plus a tetrahedron in the middle. The regions contiguous

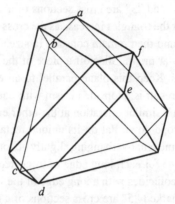

Fig. 5.10. Image of a 10-dimensional orbit; skew-cubeoctahedron.

to the small triangular faces are triangular pyramids, the regions contiguous to
large triangular faces are truncated triangular pyramids, and, finally, the regions
contiguous to the rectangular faces are triangular prismlike polyhedrons that
we have encountered earlier.

The weak-coupling techniques of Section 4.7 apply to all regions. The reduced spaces are $\mathbb{C}P^2$s. The D–H polynomial is constant in the middle tetrahedron. It is quadratic in the 14 boundary regions. The level sets of the corresponding polynomials are planes parallel to the faces and they are increasing
in the direction of the center.

A generic orbit of $SU(4)$ is 12-dimensional. Its image is depicted on Figure 5.11. The singular polygons are hexagons and rectangles. They partition
the polytope into over 100 regions. We won't attempt to describe these regions
since we know of no reasonable way of doing it.

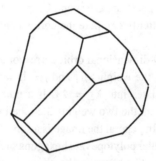

Fig. 5.11. Image of a generic $SU(4)$ orbit.

5.4 Duistermaat–Heckman Polynomials for a 10-Dimensional Orbit of $SU(4)$

Recall from Chapter 3 that the "Heckman" formula (3.50) for the Duistermaat–Heckman measure requires only the knowledge of the isotropy representations at the fixed points. Thus our first step in computing the Duistermaat–Heckman polynomials for a given orbit is to compute the moment map for the isotropy representations. Our tactic again is to go back and forth between the computations for an arbitrary compact group K and the computation for $SU(4)$.

So let K again be a compact group, λ an element of the maximal torus $\mathfrak{t}$ of $\mathfrak{k}$, and $\mathcal{O}_\lambda$ the (co)adjoint orbit of K through λ. Let $H = K_\lambda$ denote the isotropy group of λ and $\Phi = \Phi_\lambda \colon \mathcal{O}_\lambda \to \mathfrak{t}$ the moment map for the action of the maximal torus T on the orbit. A choice of an inner product on $\mathfrak{k}$ fixes an inner product on $\mathfrak{t}$ and thus fixes a choice of the Lebesgue measure on the maximal torus $\mathfrak{t}$. Let $\Psi = \Psi_\lambda$ denote the moment map for the isotropy representation of T on the tangent space T_λ to the orbit $\mathcal{O}_\lambda$ at the point λ. We will now compute the moment map Ψ. By the equivariant Darboux theorem there is a neighborhood U of λ such that for any $x \in U$ we have $\Psi(x) = \Phi(x) - \lambda$. But first we have to compute the symplectic form on the tangent space T_λ.

For $\xi \in \mathfrak{k}$ let $\xi^{\#}$ denote the vector field on the orbit induced by the infinitesimal action of ξ,

$$\xi^{\#}(x) = \left.\frac{d}{dt}\right|_{t=0} \mathrm{Ad}(\exp t\xi)x = [\xi, x]. \tag{5.10}$$

Since the orbit is a homogeneous K-space, the tangent space at λ is spanned by the vectors of the form $\xi^{\#}(\lambda)$, $\xi \in \mathfrak{k}$, that is,

$$T_\lambda = \{[\xi, \lambda] \mid \xi \in \mathfrak{k}\} = \mathrm{im}(\mathrm{ad}\,\lambda). \tag{5.11}$$

Since the orbit is embedded in the Lie algebra $\mathfrak{k}$, it makes sense to identify the tangent space with a subspace of $\mathfrak{k}$. The invariance of the inner product on $\mathfrak{k}$ implies that ker ad λ and im ad λ are mutually orthogonal and that

$$\ker(\mathrm{ad}\,\lambda) = \mathfrak{t} \oplus \sum_{\alpha \in \Delta^+, \, \alpha(\lambda)=0} Z_\alpha,$$

where Z_αs are the root spaces of K (cf. (5.6)). Therefore the tangent space to the orbit at λ is

$$T_\lambda = \sum_{\alpha \in \Delta^+, \, \alpha(\lambda) \neq 0} Z_\alpha. \tag{5.12}$$

The kernel ker(ad λ) is of course the Lie algebra of the isotropy group H of λ.

Example 5.1 Let $\mathcal{O}_\lambda$ be the orbit of $SU(4)$ through $\lambda = (3a - 2b, 2b - a, -a, -a)$, $a > b > 0$. Then $e_{ij}(\lambda) \neq 0$ if and only if $(i, j) = (1, 2), (1, 3),$ $(1, 4), (2, 3), (2, 4)$. This agrees with the fact that the isotropy group H of such a λ is the collection of matrices

$$H = \left\{ \begin{pmatrix} * & 0 & & 0 \\ 0 & * & & \\ & & * & * \\ 0 & & * & * \end{pmatrix} \right\}.$$

By construction the Z_αs $(\alpha(\lambda) \neq 0)$ are the weight spaces for the isotropy representation of T on the tangent space at λ. Moreover, we claim that these subspaces are symplectic.

Indeed, the Kirillov–Kostant–Souriau symplectic form ω on $\mathcal{O}_\lambda$ is defined on the tangent space T_λ by

$$\omega_\lambda(\xi^\#(\lambda), \eta^\#(\lambda)) = (\lambda, [\xi, \eta]) \tag{5.13}$$

for all $\xi, \eta \in \mathfrak{k}$ (cf. (5.10)). Now take $\xi = u_\alpha$ and $\eta = v_\alpha$, the basis vectors of the root space Z_α (cf. 5.5). Then (5.13) becomes

$$(\lambda, [u_\alpha, v_\alpha]) = \omega_\lambda([u_\alpha, \lambda], [v_\alpha, \lambda]). \tag{5.14}$$

Therefore, by (5.5),

$$(\lambda, (u_\alpha, u_\alpha) \cdot \alpha) = \alpha(\lambda)(-\alpha(\lambda))\omega_\lambda(v_\alpha, -u_\alpha) = (\alpha, \lambda)^2 \omega_\lambda(u_\alpha, v_\alpha).$$

Consequently

$$\omega_\lambda(u_\alpha, v_\alpha) = \frac{(u_\alpha, u_\alpha)}{(\alpha, \lambda)} \neq 0, \tag{5.15}$$

that is, Z_α is symplectic. If we assume (which we may) that λ lies in the closure of the positive Weyl chamber then $\omega_\lambda(u_\alpha, v_\alpha) > 0$.

For $\alpha \neq \beta$ the spaces Z_α and Z_β are symplectically orthogonal to each other. Here is one way to see it. Let $x \in Z_\alpha$ and $y \in Z_\beta$. Then a computation just like the one before shows that

$$(\lambda, x)(\lambda, y)\omega_\lambda(x, y) = (\lambda, [x, y]).$$

Since the root space decomposition is orthogonal, $\omega_\lambda(x, y) = 0$ if and only if $[x, y] \in \sum_{\gamma \in \Delta^+} Z_\gamma$. Now

$$[x, y] \in Z_{\alpha+\beta} \oplus Z_{|\alpha-\beta|}.$$

(Here $|\gamma|$ denotes the *positive* root collinear with a root γ.) Thus $[x, y]$ projects nontrivially onto $\mathfrak{t}$ only if $\alpha \pm \beta = 0$, which is impossible since α and β are both positive and $\alpha \neq \beta$.

Alternatively one can define a T-invariant complex structure J on T_λ by setting $Ju_\alpha = v_\alpha$ and $Jv_\alpha = -u_\alpha$. Then

$$h(x, y) = \omega_\lambda(x, Jy) + \sqrt{-1}\omega_\lambda(x, y)$$

is a T-invariant Hermitian inner product on the tangent space. The spaces Z_α become complex lines invariant under T. Since no two positive roots are collinear, these lines must be orthogonal to each other with respect to h. We conclude that the decomposition (5.12) of the tangent space is a *symplectic* direct sum.

This observation allows us to compute the moment map Ψ for the action of the torus T on the tangent space T_λ rather easily. Indeed, it is enough to compute the restriction of Ψ to each subspace Z_α and then sum up the results. Recall that if a group G acts linearly on a symplectic vector space (V, ω) preserving ω, then the ζ-component of the moment map for a vector ζ in the Lie algebra of G is given by the formula

$$\langle \zeta, \Phi(v) \rangle = \frac{1}{2}\omega(\zeta \cdot v, v),$$

where $v \in V$ and ζ on the right is thought of as an element of $\mathfrak{gl}(V)$. In our case we write an element z_α of Z_α as $a_\alpha u_\alpha + b_\alpha v_\alpha$ for some $a_\alpha, b_\alpha \in \mathbb{R}$ and compute

$$
\begin{aligned}
\frac{1}{2}\omega_\lambda(\xi \cdot z_\alpha, z_\alpha) &= \frac{1}{2}\omega_\lambda([\xi, a_\alpha u_\alpha + b_\alpha v_\alpha], a_\alpha u_\alpha + b_\alpha v_\alpha) \\
&= \frac{1}{2}\omega_\lambda(\alpha(\xi)a_\alpha v_\alpha - \alpha(\xi)b_\alpha u_\alpha, a_\alpha u_\alpha + b_\alpha v_\alpha) \\
&= \frac{1}{2}\alpha(\xi)(a_\alpha^2\omega_\lambda(v_\alpha, u_\alpha) - b_\alpha^2\omega_\lambda(u_\alpha, v_\alpha)) \\
&= -\frac{1}{2}\omega_\lambda(u_\alpha, v_\alpha)(a_\alpha^2 + b_\alpha^2)\alpha(\xi) \\
&= -\frac{1}{2}(\omega_\lambda(u_\alpha, v_\alpha)|z_\alpha|^2)\alpha(\xi).
\end{aligned}
$$

Combining with (5.15) we get

$$\Psi\left(\sum_{\alpha \in \Delta^+, \alpha(\lambda) \neq 0} z_\alpha\right) = -\sum_\alpha \frac{|u_\alpha|^2}{2(\alpha, \lambda)}|z_\alpha|^2\alpha.$$

Thus if $\alpha_1, \ldots, \alpha_N$ are the positive roots with $\alpha_i(\lambda) \neq 0$, then in the coordinates $\{a_i, b_i\}$ on $T_\lambda\mathcal{O}_\lambda$ corresponding to the basis

$$u_{i,\lambda} = \frac{(\alpha_i, \lambda)^{1/2}}{\sqrt{2}|u_\alpha|}u_{\alpha_i}, \qquad v_{i,\lambda} = \frac{(\alpha_i, \lambda)^{1/2}}{\sqrt{2}|v_\alpha|}v_{\alpha_i}$$

we have $\Psi(a_1, b_1, \ldots, a_N, b_N) = -\sum_{i=1}^{N}(a_i^2 + b_i^2)\alpha_i$. And the moment map Φ in the corresponding coordinates on the orbit $\mathcal{O}_\lambda$ in a neighborhood of λ is given by

$$\Phi(a_1, b_1, \ldots, a_N, b_N) = \lambda - \sum_{i=1}^{N}(a_i^2 + b_i^2)\alpha_i. \tag{5.16}$$

Notation

(a) For a vector v in a vector space V let v^+ denote the push-forward of the measure dt^+ by the embedding $\mathbb{R} \to V, t \to tv$.

(b) Let vol_λ denote the Liouville measure on the orbit $\mathcal{O}_\lambda$.

With this notation in mind we have *near λ in* $\mathfrak{t}^*$

$$\Phi_*(vol_\lambda) = (A_\lambda)_*(\alpha_1^+ * \cdots * \alpha_N^+), \tag{5.17}$$

where $A_\lambda : \mathfrak{t}^* \to \mathfrak{t}^*$ sends y to $\lambda - y$. Equation (5.17) together with a few simple observations is enough to compute the D–H measure for 10-dimensional orbits of $SU(4)$. We shall concentrate on the family of orbits with the spectrum of the form $\lambda = (3a - 2b, 2b - a, -a, -a)$ and leave the other case as an exercise.

We have seen in Example 5.1 that $e_{ij}(\lambda) \neq 0$ if and only if $(i, j) \in S = \{(1, 2), (1, 3), (1, 4), (2, 3), (2, 4)\}$. In the coordinates (5.9) for the maximal torus of $\mathfrak{su}(4)$

$$e_{12} = (1, 1, 0), e_{13} = (0, 1, 1), e_{14} = (1, 0, 1),$$
$$e_{23} = (-1, 0, 1), e_{24} = (0, -1, 1). \tag{5.18}$$

Thus to compute the D–H measure near λ we need to compute the convolution $e_{12}^+ * \cdots * e_{24}^+$ and then translate it by A_λ. We employ the algorithm used to prove Theorem 3.2.9 to compute the convolution in question.

Recall that if e is a vector in $\mathbb{R}^n$, A a convex closed subset of $\mathbb{R}^n$, χ_A the characteristic function of A, and g a continuous function, then the measure $e^+ * (\chi_A g)$ is absolutely continuous with respect to the Lebesgue measure dx and the Radon–Nikodym derivative f of the measure satisfies the differential equation $D_e f = \chi_A g$, that is,

$$\left.\frac{d}{dt}\right|_0 f(x + te) = \chi_A g(x).$$

To compute the function f at a point x we have to consider two cases:

1. The affine line $x + e\mathbb{R}$ intersects the set A in a point or doesn't intersect A at all. Then $f(x) = 0$.

2. The affine line intersects A in a line segment $[t_0, t_1]e + x$ (where t_0 and t_1 of course depend on x).
   Then

$$f(x + te) = \begin{cases} 0, & t \le t_0, \\ \int_{t_0}^{t} g(x + se)ds, & t_0 < t < t_1, \\ \int_{t_0}^{t_1} g(x + se)ds, & t_1 < t. \end{cases}$$

So if $x \in A$ then $f(x) = \int_{t_0}^{0} g(x+se)ds$, if $x \notin A$ and $0 < t_0$ then $f(x) = 0$, and if $x \notin A$ and $t_1 \le 0$ then $f(x) = \int_{t_0}^{t_1} g(x + se)ds$.

Example 5.2 Fix $a \ge 0$, let $A = \{(y_1, y_2) \in \mathbb{R}^2 \mid y_2 \ge a - y_1, \quad y_1 \le 0\}$, $g(y_1, y - 2) = 1$, and $e = (1, 1)$. If $y_2 \ge a + y_1$ then $((y_1, y_2) + \mathbb{R}e) \cap A$ is empty or a point, so $f(y_1, y_2) = 0$. Also if $y_2 \le a - y_1$ then $f(y_1, y_2) = 0$ as well. Suppose now that $y_2 > |y_1| + a$. Then

$$\begin{cases} y_2 + t_0 = a - y_1 - t_0, \\ y_1 + t_1 = 0. \end{cases}$$

So

$$\begin{cases} t_0 = 1/2(a - (y_1 + y_2)), \\ t_1 = -y_1. \end{cases}$$

Consequently, if $(y_1, y_2) \in A$ then

$$f(y_1, y_2) = \int_{t_0}^{t_1} ds = \frac{1}{2}(y_1 + y_2 - a),$$

and if $y_1 \ge 0$, $y_2 \ge y_1 + a$ then

$$f(y_1, y_2) = \int_{t_0}^{t_1} ds = \frac{1}{2}(-y_1 + y_2 - a).$$

Example 5.3 Let $A \subset \mathbb{R}^3$ be the cone spanned by the vectors $(0, -1, 1)$, $(1, 0, 1)$, and $(-1, 0, 1)$, $g = 1/2$, and $e = (0, 1, 1)$. Clearly the set-up is symmetric relative to the $y_1 y_2$-plane. So we compute the convolution f in the half-space $\{y_1 \ge 0\}$ and then reflect. By restricting f to the affine plane $y_1 = a$, which is parallel to e, we reduce the computation to the one carried out in the previous example with y_2 and y_3 playing the roles of y_1 and y_2, and with $g = 1/2$. Thus for y in the cone spanned by $(1, 0, 1)$, $(0, -1, 1)$, and $(0, 0, 1)$

$$f(y) = 1/4(-y_1 + y_2 + y_3),$$

and for y in the cone spanned by $(1, 0, 1)$, $(0, 1, 1)$, and $(0, 0, 1)$

$$f(y) = 1/4(-y_1 - y_2 + y_3).$$

By reflecting f across the plane $\{y_1 = 0\}$ we get $f(y) = 1/4(y_1 + y_2 + y_3)$ and $f(y) = (y_1 - y_2 + y_3)$ in the appropriate cones. Of course, f is zero outside the four cones.

Note, by the way, that

$$f \, dy = e_{13}^+ * \cdots * e_{24}^+$$

(cf. (5.18)). So in order to compute the convolution $h \, dy = e_{12}^+ * \cdots * e_{24}^+$ it is enough to compute the convolution $e_{12}^+ * f(y) \, dy$, where $f \, dy$ is the measure just computed. This is what we will do presently. It is easy to see that the measure $h \, dy$ is supported on the cone spanned by the vectors $(0, -1, -1)$, $(-1, 0, 1)$, and $(1, 1, 0)$ and that h is polynomial in the cones defined by the following inequalities:

(1) $y_1, y_2 \leq 0$, $y_1 + y_2 + y_3 \geq 0$
    (a) $y_1 \leq 0, y_2 \geq 0$, $y_1 - y_2 + y_3 \geq 0$
    (b) $y_1 \geq 0$, $y_2 \leq 0$, $-y_1 + y_2 + y_3 \geq 0$
(2) $|y_1 - y_2| \leq y_3$, $y_1 + y_2 - y_3 \geq 0$
(3) $y_1, y_2 \geq 0$, $-y_1 - y_2 + y_3 \geq 0$.

(The reason for the idiosyncratic notation may become clear later.) Once again we compute the convolution by restricting the measure to an affine plane parallel to $e = e_{12}$, in this case to $\{y_3 = a\}$. The computation itself is very similar to the one carried out in Example 5.2, so we simply list the results. Let us denote the restriction of h to region 1 by h_1, to region a by h_a, and so forth. Then

$$
\begin{aligned}
h_1(y) &= \tfrac{1}{16}(y_1 + y_2 + y_3)^2, \\
h_a(y) &= \tfrac{1}{16}((y_1 + y_2 + y_3)^2 - 4y_2^2), \\
h_b(y) &= \tfrac{1}{16}((y_1 + y_2 + y_3)^2 - 4y_2^2), \qquad (5.19)\\
h_3(y) &= \tfrac{1}{16}((y_1 + y_2 + y_3)^2 - 4y_1^2 - 4y_2^2), \\
h_4(y) &= \tfrac{1}{16}2(y_3^2 - (y_1 - y_2)^2).
\end{aligned}
$$

Let us also record the jumps in the polynomials for future reference:

$$
\begin{aligned}
(h_1 - h_a)(y) &= -\tfrac{4}{16}y_2^2, \\
(h_4 - h_3)(y) &= -\tfrac{1}{16}(y_1 + y_2 - y_3)^2.
\end{aligned}
\qquad (5.20)
$$

This concludes the preliminary computations. We are now in the position to write down the D–H polynomials on a 10-dimensional orbit.

Combining (5.17) and (5.19) we can now write down the measure $\Phi_*(vol_\lambda)$ at the points y near $\lambda = (a - b, a - b, a)$ and therefore on all the components of regular values of Φ that have λ as a vertex. Before we do that, we need a simple way to keep track of 40-odd regions of regular values. Let us suppose again that λ is close to the line through the fundamental weight μ_3 (cf. Figure 5.2), that is, that $b << a$. Now look at the cross section of the image polytope by the plane $\{y_1 = y_2\}$. The result is depicted in Figures 5.12 and 5.13 for $b = 2/5 \, a$. The cross sections 1–7 that we have employed in the previous section are orthogonal to the plane $\{y_1 = y_2\}$. For the reader's convenience their intersections with the plane are also marked in Figure 5.12, and the regions Δ_2–Δ_6 on the cross sections are numbered in Figure 5.13 in the same way. Due to the symmetry considerations, it is enough to write down the D–H polynomials in regions Δ_1–Δ_7.

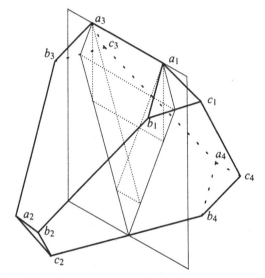

Fig. 5.12. Slicing the image of a 10-dimensional orbit.

On Δ_1 we have $\Phi_*(vol_\lambda) = f_1(y) \, dy$, where

$$f_1(y) = h_1(\lambda - y) = \frac{1}{16}(3a - 2b - y_1 - y_2 - y_3)^2.$$

Similarly on Δ_4

$$f_4(y) = \frac{1}{16}((3a - 2b - y_1 - y_2 - y_3)^2 - 4(y_1 - a + b)^2 - 4(y_2 - a + b)^2)$$

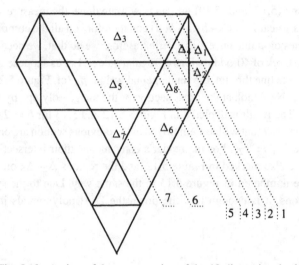

Fig. 5.13. A view of the cross section of the 10-dimensional orbit.

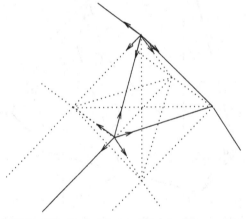

Fig. 5.14. Some weights entering in the computation of the measure associated with the 10-dimensional orbit.

and on Δ_3

$$f_3(y) = \frac{1}{16}2((y_3 - a)^2 - (y_1 - y_2)^2).$$

To compute the remaining four polynomials we use Weyl group symmetry and the following observation.

Proposition 5.4.1 *The jump polynomial along a Delzant (completely integrable) wall is independent of the point of the jump.*

Proof. This is just a restatement of the discussion in Section 3.5, (3.90) in particular. Alternatively one can use the "Heckman" formula (3.50) to establish the proposition. We will return to this point later.   □

Proposition 5.4.1 asserts that

$$f_1 - f_2 = f_4 - f_8 = f_3 - f_5$$

and that

$$f_3 - f_4 = f_5 - f_8 = f_7 - f_8.$$

A word of caution: One might be tempted to think that the pairs of regions Δ_1 and Δ_4, Δ_2 and Δ_8, Δ_5 and Δ_7, and Δ_8 and Δ_6 share a common face. This is not true – they only share an edge – and so Proposition 5.4.1 does not apply directly to these pairs of regions.

We now compute $f_1 - f_2$ and f_2. The regions Δ_a and Δ_b corresponding to the polynomials h_a and h_b intersect the plane $\{y_1 = y_2\}$ in line segments, and thus do not appear in Figure 5.13. However the reflection $s_{23}(y_1, y_2, y_3) = (y_1, y_3, y_2)$ sends Δ_a to Δ_2. Thus

$$(f_1 - f_2)(y) = (f_1 - f_a) \circ s_{23}(y) = (h_1 - h_a) \circ s_{23}(\lambda - y) = -\frac{4}{16}(a - b - y_3)^2$$

and

$$f_2(y) = \frac{1}{16}((3a - 2b - y_1 - y_2 - y_3)^2 - 4(y_3 - a + b)^2).$$

Consequently,

$$f_5(y) = (f_3 - (f_1 - f_2))(y) = \frac{1}{16}2((y_3 - a)^2 - (y_1 - y_2)^2 - 2(a - b - y_3)^2)$$

and

$$
\begin{aligned}
f_8(y) &= (f_4 - (f_1 - f_2))(y) \\
&= \frac{1}{16}(((3a - 2b - y_1 - y_2 - y_3)^2 - 4(y_1 - a + b)^2 \\
&\quad - 4(y_2 - a + b)^2) - 4(y_3 - a + b)^2).
\end{aligned}
$$

Note that f_8 is invariant under the permutations of y_1, y_2, and y_3, which are the elements of the Weyl group that preserve the weight μ_3. Note also that we can write f_8 as

$$f_1 - ((f_1 - f_2) + (f_1 - f_a) + (f_1 - f_b)).$$

This is because the regions Δ_8 and Δ_1 share a vertex and so one can get from Δ_1 to Δ_8 by crossing the three walls locally given by $\{y_1 = a - b\}$, $\{y_2 = a - b\}$, and $\{y_3 = a - b\}$. The regions Δ_8 and Δ_7 also have a vertex in common. The three walls separating the two regions are the wall locally given by

$$y \cdot (-1, -1, -1) = (a - b, a - b, a) \cdot (-1, -1, -1)$$

(the wall that separates Δ_3 and Δ_4) and its two images under the reflections s_{23} and s_{13}: $(y_1, y_2, y_3) \mapsto (y_3, y_2, y_1)$. So

$$f_7 = f_8 + (f_3 - f_4) + (f_3 - f_4) \circ s_{13} + (f_3 - f_4) \circ s_{23}.$$

Now

$$(f_3 - f_4)(y) = (h_3 - h_4)(\lambda - y) = \frac{1}{16}(a - 2b - y_1 - y_2 + y_3)^2.$$

Collecting the terms we get

$$f_7(y) = \frac{1}{16}4b^2.$$

Thus Δ_7 is a lacunary region (cf. Theorem 4.5.1 and the subsequent discussion). Finally observe that

$$f_7 - f_6 = f_3 - f_4,$$

so

$$f_6(y) = \frac{1}{16}(4b^2 - (a - 2b - y_1 - y_2 + y_3)^2).$$

A natural question now arises: What is the range of λs for which the preceding computation is valid? Moving the point $\lambda = (a - b, a - b, a)$ along the line through the origin simply rescales the diagram. So the one interesting parameter is the ratio b/a (in Figure 5.12 $b/a = 2/5$). Let us fix for the moment the value of a. If we let $b \to 0$, the region Δ_7 fills out the whole diagram, as we would expect. But *no region disappears* until b is actually zero, and then we are looking at the image of a six-dimensional orbit over which our family of orbits fibers. What if we let b increase from $2/5\, a$? Then Δ_7 shrinks down, and it disappears when the triangular walls pass through the origin. This occurs precisely when

$$(-1, -1, -1) \cdot (a - b, a - b, a) = 0,$$

or $a/2 = b$. Thus our computation is valid for $0 < b < a/2$. In particular the measure in the central region Δ_7 is constant whenever $0 < b < a/2$, which is a rather large range of parameters.

This phenomenon is not really all that surprising in view of the fact that the polynomials f_i depend piecewise polynomially on the coordinates of λ. In the next section we will give two proofs of this fact, one based on minimal coupling and reduction and one based on the "Heckman" formula (3.50) and on equation (5.17).

5.5 Variation of Orbits

As before let $\mathcal{O}_\lambda$ be an adjoint orbit of a compact group K passing through the point λ of the maximal torus $\mathfrak{t}$, and let $\Phi_\lambda \colon \mathcal{O}_\lambda \to \mathfrak{t}$ be the moment map. For a regular value β of Φ_λ denote the reduced space at β by $M(\beta, \lambda)$. The point of introducing more notation is that we would now like to study the dependence of the moment map on the point $\lambda \in \mathfrak{t}$ and also the dependence on λ of the reduced spaces and of the corresponding D–H (interpolating) polynomials.

Let us start by comparing the reduced spaces $M(\beta, \lambda)$ as β and λ vary while keeping the diffeotype of the orbit $\mathcal{O}_\lambda$ fixed. For this purpose fix the Weyl chamber wall $\mathfrak{u}$ containing the point λ in its interior (if λ is a regular point of $\mathfrak{t}$ then $\mathfrak{u} = \mathfrak{t}$). Let U be the corresponding subgroup of T. The group U is (the identity component of) the center of the stabilizer of λ.

Theorem 5.5.1 *There exists a Hamiltonian $(T \times U)$-space Y with moment map $J \colon Y \to \mathfrak{t} \times \mathfrak{u}$ such that if λ lies in the interior of the wall $\mathfrak{u}$ and β is a regular value of $\Phi_\lambda \colon \mathcal{O}_\lambda \to \mathfrak{t}$ then (β, λ) is a regular value of J. Moreover, the reduced space at (β, λ) for the action of $T \times U$ on Y is exactly $M(\beta, \lambda)$ (as a symplectic manifold):*

$$J^{-1}(\beta, \lambda)/T \times U = \Phi_\lambda^{-1}(\beta)/T.$$

Consequently, the Duistermaat–Heckman theorem applies and the coefficients of D–H polynomials associated with the map Φ_λ are piecewise polynomial in λ.

Proof. The construction is quite simple. Let us assume first to avoid technicalities that λ is a regular point of $\mathfrak{t}$, that is, that $\mathfrak{u} = \mathfrak{t}$. Recall from Section 2.3 that a choice of a left-invariant connection A on the principal T-bundle $K \to \mathcal{O}_\lambda$ gives rise to a presymplectic form $\Omega = -d\langle pr_2, A \rangle$ on $K \times \mathfrak{u}$ which is symplectic on $K \times \mathfrak{u}_0$. The torus T acts on K by multiplication on the left and by multiplication on the right by the inverse:

$$T \times T \times K \to K, \quad (a, b, g) \mapsto agb^{-1}.$$

The resulting action of $T \times T$ on $Y := K \times \mathfrak{u}_0$ (the action on $\mathfrak{u}_0$ being trivial) is Hamiltonian with moment map $J \colon K \times \mathfrak{u}_0 \to \mathfrak{t} \times \mathfrak{t}$ being given by

$$(g, \mu) \mapsto (\mathrm{Ad}(g)\mu, -\mu).$$

If we reduce Y with respect to to the action of the second torus at a point $\mu \in \mathfrak{u}_0$ the result is $\mathcal{O}_\mu$, the K-orbit through μ. The action of the first torus then descends to the standard action of T on the orbit. We conclude that if (β, μ) is a regular value of J then the reduced space $J^{-1}(\beta, \mu)/(T \times T)$ is symplectically diffeomorphic to $M(\beta, \mu) = \Phi_\mu^{-1}(\beta)/T$.

In general the stabilizer of λ is not abelian, but rather a product of a semisimple group H with U. Now, the action of $H \times U \times T$ on $K \times \mathfrak{h} \times \mathfrak{u}_0 \times \mathfrak{t}$ is Hamiltonian and, as before, the reduction at $(0, \beta, \lambda)$ gives us $M(\beta, \lambda)$. However, to apply the Duistermaat–Heckman theorem we must obtain the space $M(\beta, \lambda)$ as a result of an *abelian* reduction. So we reduce $K \times \mathfrak{h} \times \mathfrak{u}_0 \times \mathfrak{t}$ at 0 with respect to the action of H alone. This gives us a Hamiltonian $(T \times U)$-space $Y := (K/H) \times \mathfrak{u}_0$. $\square$

Example 5.4 (Variation of a generic $SU(3)$ orbit) A Weyl chamber C_0 of $SU(3)$ is a cone in $\mathbb{R}^2$ and the image of a generic orbit $\mathcal{O}_\lambda$ is a semiregular hexagon:

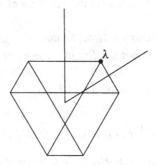

When the point λ moves along a radial line, the image of the orbit and the corresponding measure get rescaled. Let us see what happens when λ moves from the proximity of one wall to another:

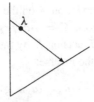

In terms of Theorem 5.5.1, it means doing the following thing. First we consider the space $Y = SU(3) \times C_0$. The image of Y under the moment map corresponding to the action of the four-torus $T \times T$ (T being a maximal torus of $SU(3)$) is a polyhedral cone in $\mathbb{R}^4$ spanned by a three-dimensional polytope. Up to some scaling this polytope is an octahedron, which supports a piecewise-linear measure dx (cf. Section 5.3). Strictly speaking we should delete two faces, one opposite another. These two triangles correspond to the points on the boundary of C_0. A cross section of the octahedron by an affine plane parallel to the deleted faces is a semiregular hexagon. When the plane is close to one face the cross section is almost a triangle. As the plane moves closer to the center the cross section approaches a regular hexagon. It becomes a regular hexagon when the plane passes through the center, and then deforms back to a triangle.

The singular squares of the octahedron break up each cross section into seven regions. The restriction of dx to such a region is a linear measure. Notice that it is constant in the middle region.

The fact that this example works out so nicely is due to a coincidence. Consider the orbit $\mathcal{O}$ of $SU(4)$ diffeomorphic to a Grassmannian of 2-planes in $\mathbb{C}^4$. The circle subgroup of $SU(4)$ generated by the weight μ_1 (in the notation of Section 5.3) acts on $\mathcal{O}$ in a Hamiltonian fashion and the corresponding reduced spaces are generic orbits of $SU(3)$.

We now present a different proof of piecewise-polynomial dependence of the measure $(\Phi_\lambda)_*(vol_\lambda)$ on λ, which is based on the "Heckman" formula and on (5.17).

Note first that the right-hand side of (5.17) is simply the summand v_p in (3.50) for $p = \lambda$. The other summands are easy to compute as well. If $q \in \mathcal{O}_\lambda$ is fixed by the torus T, then $q = w \cdot \lambda$ for some w in the Weyl group W. The renormalized weights at q are then given by

$$\alpha_i^w = \epsilon_i(w \cdot \alpha_i),$$

where $\epsilon_i = +1$ or -1 depending on whether $w \cdot \alpha_i$ is a positive or a negative weight (cf. (3.49)). Arguing exactly the same way as we did to derive (5.17), we get

$$v_q = (A_{w \cdot \lambda})_*((\alpha_1^w)^+ * \cdots * (\alpha_N^w)^+),$$

where $A_{w \cdot \lambda}(y) = w \cdot \lambda - y$. We conclude that for a coadjoint orbit $\mathcal{O}_\lambda$ the "Heckman" formula (3.50) simply becomes

$$(\Phi_\lambda)_*(vol_\lambda) = \frac{1}{|W_\lambda|} \sum_{w \in W} (-1)^{\ell(w)} (A_{w \cdot \lambda})_*((\alpha_1^w)^+ * \cdots * (\alpha_N^w)^+), \quad (5.21)$$

where $|W_\lambda|$ is the cardinality of the isotropy group of λ in W and $\ell(w)$ is the length of w. Now for any other point λ' in the interior of the wall $\mathfrak{u}$ and any positive root α

$$\alpha(\lambda') \neq 0 \iff \alpha(\lambda) \neq 0.$$

Thus the weights α_i^w in (5.21) *do not depend on* λ; they only depend on the wall $\mathfrak{u}$. Therefore (5.21) is valid for all λ in the interior of $\mathfrak{u}$ with the weights α_i^w being fixed. Clearly the Radon–Nikodym derivative $f_\lambda(y)$ of the right-hand side of (5.21) is a piecewise-polynomial function of λ.

Appendix A: Multiplicity Formulas

A.1 Weyl, Kostant, and Steinberg Formulas

Since the fundamental work of Bernstein, Gelfand, and Gelfand [BGG] on Verma modules, and a basic observation by Kac (cf., for example, [Kac]), the proofs of the fundamental multiplicity formulas can be derived in an algebraic way by very elementary arguments. For the convenience of the reader we present these proofs here, even though they can be found quite readily in the literature. We follow the treatment given in Humphreys [Hum] and for that reason, in this appendix, we will adhere with some deviations to his notation.

Notation. L denotes a semisimple Lie algebra with Killing form κ. We choose a Cartan subalgebra H. The set of all roots is denoted by Φ and we choose a set of positive roots Φ^+ and denote the associated system of simple roots by Δ.

The Killing form gives an identification of H^* with H,

$$H^* \ni \lambda \mapsto t_\lambda \in H, \quad \kappa(t_\lambda, h) = \lambda(h) \quad \forall h \in H.$$

The induced bilinear form on H^* is denoted by $(\ ,\ )$ so

$$(\lambda, \phi) := \kappa(t_\lambda, t_\phi).$$

The root spaces are denoted by L_α, so $\kappa(L_\alpha, L_\beta) = 0$ if $\beta \neq -\alpha$ and $\kappa(L_\alpha, H) = 0$, and therefore L_α is nonsingularly paired with $L_{-\alpha}$ by the Killing form. Furthermore, for $x \in L_\alpha, y \in L_{-\alpha}$ we have

$$[x, y] = \kappa(x, y)t_\alpha. \tag{A.1}$$

This follows immediately from the invariance of the Killing form. Indeed, for any $h \in H$ we have

$$
\begin{aligned}
\kappa(h, [x, y]) &= \kappa([h, x], y) \\
&= \alpha(h)\kappa(x, y) \\
&= \kappa(h, t_\alpha)\kappa(x, y) \\
&= \kappa(h, \kappa(x, y)t_\alpha).
\end{aligned}
$$

The nondegeneracy of κ restricted to H thus implies (A.1).

For $\lambda \in H^*$, $V(\lambda)$ denotes the unique irreducible module of highest weight λ and $M(\lambda)$ denotes the Verma module of highest weight λ. So $M(\lambda)$ is the "largest" cyclic highest-weight module with highest weight λ. More generally, $Z(\lambda)$ denotes an arbitrary cyclic module of highest weight λ.

The existence and uniqueness of the Verma module is an immediate consequence of the Poincaré–Birkhoff–Witt theorem, as is the fact that a basis for $M(\lambda)$ consists of all vectors of the form

$$
y_{\beta_1}^{i_1} \cdots y_{\beta_N}^{i_N} v,
$$

where a nonzero $y_\beta \in L_{-\beta}$ is chosen (once and for all) for each $\beta \in \Phi^+$. For any $\lambda \in H^*$ the module $M(\lambda)$ has a maximal proper submodule and hence a unique irreducible quotient which we have denoted by $V(\lambda)$.

The **weight lattice** $\Lambda \subset H^*$ consists of those λ that satisfy

$$
\langle \lambda, \alpha \rangle \in \mathbb{Z}
$$

for all weights α, where

$$
\langle \lambda, \alpha \rangle := \frac{2(\lambda, \alpha)}{(\alpha, \alpha)}.
$$

The irreducible representation $V(\lambda)$ is finite-dimensional if and only if $\lambda \in \Lambda^+$, the set of **dominant weights**, where Λ^+ is the intersection of Λ with the closure of the positive Weyl chamber, that is, it is the set of all $\lambda \in \Lambda$ that satisfy

$$
\langle \lambda, \alpha \rangle \geq 0, \quad \forall \alpha \in \Delta.
$$

This is all given in chapter III and the first few sections of chapter VI in Humphreys.

We set

$$
\delta := \frac{1}{2} \sum_{\beta \in \Phi^+} \beta = \lambda_1 + \cdots + \lambda_n,
$$

where the λ_i form the basis of the root lattice Λ dual under $\langle \, , \, \rangle$ to the simple roots.

The **Kostant partition function** $P_K(\mu)$ is defined as the number of sets of nonnegative integers k_β such that

$$\mu = \sum_{\beta \in \Phi^+} k_\beta \beta.$$

In other words, it is the number of ways that μ can be expressed as a sum of positive roots. (The value is zero if μ cannot be expressed as a sum of positive roots.)

For any module N, and any $\mu \in H^*$, N_μ denotes the weight space of weight μ. For example, in the Verma module $M(\lambda)$ the only nonzero weight spaces are the ones where $\mu = \lambda - \sum_{\beta \in \Phi^+} k_\beta \beta$ and the multiplicity of this weight space, that is, the dimension of $M(\lambda)_\mu$, is the number of ways of expressing μ in this fashion, that is,

$$\dim M(\lambda)_\mu = P_K(\lambda - \mu). \tag{A.2}$$

On Λ we introduce the partial order

$$\mu \prec \lambda \Leftrightarrow \lambda - \mu = \sum_{\alpha \in \Phi^+} k_\alpha \alpha,$$

where the k_α are nonnegative integers. Thus the right-hand side of (A.2) vanishes unless $\mu \prec \lambda$.

For each $\lambda \in \Lambda$ we introduce a formal symbol $e(\lambda)$ which we want to think of as an "exponential" and so the symbols are multiplied according to the rule

$$e(\mu) \cdot e(\nu) = e(\mu + \nu).$$

The **character** of a module N is defined as

$$\mathrm{ch}_N := \sum \dim N_\mu \cdot e(\mu).$$

In all cases we will consider (cyclic highest-weight modules and the like) all these dimensions will be finite, so the coefficients are well defined, but (in the case of Verma modules, for example) there may be infinitely many terms in the (formal) sum. Logically, such a formal sum is nothing other than a function on Λ giving the "coefficient" of each $e(\mu)$. So, for example, we might write the formula (A.2) for the mulitiplicities of the Verma module as

$$\mathrm{ch}_{M(\lambda)} = \sum P_K(\lambda - \mu)e(\mu) = P_K(\lambda - \cdot).$$

To be consistent with Humphreys's notation, define the **Kostant function** p by

$$p(\nu) = P_K(-\nu)$$

and then in succinct language

$$\mathrm{ch}_{M(\lambda)} = p(\cdot - \lambda). \tag{A.3}$$

The formulas we want are, for $\lambda \in \Lambda^+$:

- The **Kostant Multiplicity formula:** For any $\lambda \in \Lambda^+$, $\mu \in \Lambda$,

$$\dim V(\lambda)\mu = \sum_{w \in W} (-1)^w p(\mu + \delta - w(\lambda + \delta)), \qquad \text{(A.4)}$$

  where we have written $(-1)^w$ for $(-1)^{\ell(w)}$.

- The **Weyl dimension formula:**

$$\dim V(\lambda) = \frac{\prod_{\beta \in \Phi^+}(\lambda + \delta, \beta)}{\prod_{\beta \in \Phi^+}(\delta, \beta)}. \qquad \text{(A.5)}$$

- The **Weyl character formula:**

$$\text{ch}_{V(\lambda)} = \frac{\sum_{w \in W}(-1)^w e(w(\lambda + \delta))}{\sum_{w \in W}(-1)^w e(w\delta)}. \qquad \text{(A.6)}$$

- The **Steinberg formula** for the number of times that the irreducible $V(\lambda)$ occurs in the decomposition of the tensor product of two irreducibles: The multiplicity of $V(\lambda)$ in $V(\lambda') \otimes V(\lambda'')$ is given by

$$\sum_{w,s \in W} (-1)^{sw} p(\lambda + 2\delta - s(\lambda' + \delta) - w(\lambda'' + \delta)). \qquad \text{(A.7)}$$

We will prove the Kostant and Weyl formulas. The Steinberg formula is an immediate consequence of them (cf. [Hum], pp. 140–1).

A.2 The Action of the Casimir Element on $Z(\lambda)$

Recall that if we choose a basis $\{v_1, \ldots, v_M\}$ of the Lie algebra and the dual basis $\{w_1, \ldots, w_M\}$ relative to the Killing form κ, then the Casimir element

$$c = \sum v_i w_i$$

is independent of the choice of basis, and lies in the center of the universal enveloping algebra. In particular, it must be a scalar times the identity operator in any cyclic module. We will prove that in any module $Z(\lambda)$ this scalar is given by

$$c(\lambda) := c_{Z(\lambda)} = (\lambda + \delta, \lambda + \delta) - (\delta, \delta). \qquad \text{(A.8)}$$

To prove this, it is enough to check that $cv = c(\lambda)v$, where $c(\lambda)$ is given by the previous formula and where v is the highest-weight vector. For this we choose a convenient basis of L. Let $h_1, \ldots, h_n$ be the standard basis of H and choose (arbitrarily) a nonzero element $x_\beta \in L_\beta$ for each root β, positive or negative.

The dual basis is then given as $\{k_1, \ldots, k_n \mid z_\beta \in L_{-\beta}\}$, where z_β is determined by

$$\kappa(x_\beta, z_\beta) = 1.$$

It follows from (A.1) that we have

$$[x_\beta, z_\beta] = t_\beta \ \forall \beta \in \Phi^+. \tag{A.9}$$

Now each $L_\beta, \beta \in \Phi^+$, acts trivially on v, and hence $z_\beta v = 0$, $\forall \beta \in \Phi^-$. For $\beta \in \Phi^+$ write

$$x_\beta z_\beta = x_\beta z_\beta + [x_\beta, z_\beta]$$

so

$$x_\beta z_\beta v = \lambda(t_\beta)v = (\lambda, \beta)v.$$

On the other hand,

$$\sum h_i k_i v = \lambda(h_i)\lambda(k_i)v = (\lambda, \lambda)v,$$

since for any pair of dual bases and any linear function we have $\sum \lambda(h_i)\lambda(k_i) = (\lambda, \lambda)$. So c acts on v by the scalar

$$(\lambda, \lambda) + \sum_{\beta \in \Phi^+} (\lambda, \beta) = (\lambda, \lambda) + 2(\lambda, \delta)$$

$$= (\lambda + \delta, \lambda + \delta) - (\delta, \delta). \qquad \square$$

We now use the innocuous looking formula (A.8) to prove the following:

Proposition A.2.1 *Any cyclic highest-weight module $Z(\lambda)$, $\lambda \in \Lambda$, has a composition series whose quotients are irreducible modules $V(\mu)$, where $\mu \prec \lambda$ satisfies*

$$(\mu + \delta, \mu + \delta) = (\lambda + \delta, \lambda + \delta). \tag{A.10}$$

In fact, if

$$d = \sum \dim Z(\lambda)_\mu,$$

where the sum is over all μ satisfying (A.10), then there are at most d steps in the composition series.

Remark. There are only finitely many $\mu \in \Lambda$ satisfying (A.10) since the set of all μ satisfying (A.10) is compact and Λ is discrete. Each weight is of finite multiplicity. Therefore d is finite.

Proof (by induction on d). We first show that if $d = 1$ then $Z(\lambda)$ is irreducible. Indeed, if not, any proper submodule, being the sum of its weight spaces, must have a highest-weight vector with highest-weight μ, say. But then $c(\lambda) = c(\mu)$ since W is a submodule of $Z(\lambda)$ and c takes on the constant value $c(\lambda)$ on $Z(\lambda)$. Thus μ and λ both satisfy (A.10), contradicting the assumption $d = 1$. In general, suppose that $Z(\lambda)$ is not irreducible, so it has a submodule W and quotient module $Z(\lambda)/W$. Each of these is a cyclic highest-weight module, and we have a corresponding composition series on each weight space. In particular, $d = d_W + d_{Z(\lambda)/W}$, so the ds are strictly smaller for the submodule and the quotient module. Hence we can apply induction. $\square$

It now follows that

$$\mathrm{ch}_{Z(\lambda)} = \sum \mathrm{ch}_{V(\mu)},$$

where the sum is over the finitely many terms in the composition series. In particular, we can apply this to $Z(\lambda) = M(\lambda)$, the Verma module. Let us order the $\mu_i \prec \lambda$ satisfying (A.10) in such a way that $\mu_i \prec \mu_j \Rightarrow i \leq j$. Then for each of the finitely many μ_i occurring we get a corresponding formula for $\mathrm{ch}_{M(\mu_i)}$ and so we get a collection of equations

$$\mathrm{ch}_{M(\mu_j)} = \sum a_{ij} \, \mathrm{ch}_{V(\mu_i)},$$

where $a_{ii} = 1$ and $i \leq j$ in the sum. We can invert this upper triangular matrix and therefore know that there is a formula of the form

$$\mathrm{ch}_{V(\lambda)} = \sum b(\mu) \, \mathrm{ch}_{M(\mu)}, \qquad (A.11)$$

where the sum is over $\mu \prec \lambda$ satisfying (A.10).

A.3 Determining the Coefficients

We will now prove:

Proposition A.3.1 *The nonzero coefficients in (A.11) occur only when*

$$\mu = w(\lambda + \delta) - \delta,$$

and then

$$b(\mu) = (-1)^w.$$

We will prove this by proving some combinatorial facts about multiplication of sums of exponentials. Observe first that if

$$f = \sum f(\mu) e(\mu)$$

then

$$f \cdot e(\lambda) = \sum f(\mu)e(\lambda + \mu) = \sum f(\nu - \lambda)e(\nu).$$

We can express this by saying that

$$f \cdot e(\lambda) = f(\cdot - \lambda).$$

Thus, for example,

$$\mathrm{ch}_{M(\lambda)} = p(\cdot - \lambda) = p \cdot e(\lambda).$$

Also observe that if

$$f_\alpha = \frac{1}{1 - e(-\alpha)} := 1 + e(-\alpha) + e(-2\alpha) + \cdots$$

then

$$(1 - e(-\alpha))f_\alpha = 1$$

and

$$\prod_{\alpha \in \Phi^+} f_\alpha = p$$

by the definition of the Kostant function.

Define the function q by

$$q := \prod_{\alpha \in \Phi^+} (e(\alpha/2) - e(-\alpha/2)) = e(\delta) \prod (1 - e(-\alpha))$$

since $e(\delta) = \prod_{\alpha \in \Phi^+} e(\alpha/2)$. Notice that

$$wq = (-1)^w q.$$

Indeed, it is enough to check this on fundamental reflections, but they have the property that they make exactly one positive root negative, hence change the sign of q.

We have

$$qp = e(\delta). \tag{A.12}$$

Indeed,

$$
\begin{aligned}
qpe(-\delta) &= \left[\prod(1 - e(-\alpha))\right]e(\delta)\,pe(-\delta) \\
&= \left[\prod(1 - e(-\alpha))\right]p \\
&= \prod(1 - e(-\alpha)) \prod f_\alpha \\
&= 1.
\end{aligned}
$$

Therefore,

$$q\mathrm{ch}_M(\lambda) = qpe(\lambda) = e(\delta)e(\lambda) = e(\lambda + \delta).$$

Let us now multiply both sides of (A.11) by q and use the preceding equation. We obtain

$$q\mathrm{ch}_{V(\lambda)} = \sum c(\mu)e(\mu + \delta),$$

where the sum is over all $\mu \prec \lambda$ satisfying (A.10).

Now $\mathrm{ch}_{V(\lambda)}$ is invariant under W, and q transforms by $(-1)^w$. Hence if we apply $w \in W$ to the preceding equation we obtain

$$(-1)^w q\mathrm{ch}_{V(\lambda)} = \sum c(\mu)e(w(\mu + \delta)).$$

This shows that the set of $\mu + \delta$ with nonzero coefficients is stable under W and the coefficients transform by the sign representation for each W orbit. In particular, each element of the form $\mu = w(\lambda + \delta) - \delta$ has $(-1)^w$ as its coefficient. We can thus write

$$q\mathrm{ch}_{V(\lambda)} = \sum_{w \in W} (-1)^w e(w(\lambda + \delta)) + R,$$

where R is a sum of terms corresponding to $\mu + \delta$ that are not of the form $w(\lambda + \delta)$. We claim that there are no such terms and hence $R = 0$. Indeed, if there were such a term, the transformation properties under W would demand that there be such a term with $\mu + \delta$ in the closure of the Weyl chamber, that is, $\mu + \delta \in \Lambda^+$. But we claim that

$$\mu \prec \lambda, (\mu+\delta, \mu+\delta) = (\lambda+\delta, \lambda+\delta), \qquad \text{and} \qquad \mu+\delta \in \Lambda^+ \implies \mu = \lambda.$$

Indeed, write $\mu = \lambda - \pi$, $\pi = \sum k_\alpha \alpha$, $k_\alpha \geq 0$, so

$$
\begin{aligned}
0 &= (\lambda + \delta, \lambda + \delta) - (\mu + \delta, \mu + \delta) \\
&= (\lambda + \delta, \lambda + \delta) - (\lambda + \delta - \pi, \lambda + \delta - \pi) \\
&= (\lambda + \delta, \pi) + (\pi, \mu + \delta) \\
&\geq (\lambda + \delta, \pi) \text{ (since } \mu + \delta \in \Lambda^+) \\
&\geq 0,
\end{aligned}
$$

since $\lambda + \delta \in \Lambda^+$. But the last inequality is strict unless $\pi = 0$. Hence $\pi = 0$. So we have derived the fundamental formula

$$q\mathrm{ch}_{V(\lambda)} = \sum_{w \in W} (-1)^w e(w(\lambda + \delta)). \qquad (A.13)$$

A.4 Proof of the Kostant and Weyl Formulas

Multiply (A.13) by $pe(-\delta)$ and use that fact that $qpe(-\delta) = 1$ to get

$$
\begin{aligned}
\mathrm{ch}_{V(\lambda)} &= qpe(-\delta)\,\mathrm{ch}_{V(\lambda)} \\
&= e(-\delta)p\sum(-1)^w e(w(\lambda+\delta)) \\
&= p\sum(-1)^w e(w(\lambda+\delta)-\delta) \\
&= \sum(-1)^w pe(w(\lambda+\delta)-\delta) \\
&= \sum(-1)^w p(\cdot+\delta-w(\lambda+\delta)).
\end{aligned}
$$

By the definition of the character, this is precisely the content of the Kostant multiplicity formula.

Notice that if we take $\lambda = 0$ and so have the trivial representation with character 1 for $V(\lambda)$, (A.13) becomes

$$
q = \sum(-1)^e(w\delta)
$$

and this is precisely the denominator in the Weyl character formula. Thus (A.13) becomes the Weyl character formula.

If for any weight μ we define

$$
A_\mu := \sum_{w \in W}(-1)^w e(w\mu)
$$

then we can write the Weyl character formula as

$$
\mathrm{ch}_{V(\lambda)} = \frac{A_{\lambda+\delta}}{A_\delta}.
$$

We now derive the Weyl dimension formula from the Weyl character formula following the treatment in Fulton and Harris [FH]. For any weight μ define the homomorphism Ψ_μ from the ring of formal combinations of the symbols $e(\nu)$ into the ring of formal power series in one variable t by the formula

$$
\Psi_\mu(e(\nu)) = e^{(\nu, mu)t}
$$

(and extend linearly). We claim that

$$
\Psi_\mu(A_\nu) = \Psi_\nu(A_\mu)
$$

for any pair of weights. Indeed,

$$\Psi_\mu(A_v) = \sum_w (-1)^w e^{(\mu, wv)t}$$

$$= \sum_w (-1)^w e^{(w^{-1}v, \mu)t}$$

$$= \sum_w (-1)^w e^{(w\mu, v)t}$$

$$= \Psi_v(A_m u).$$

In particular,

$$\Psi_\delta(A_\lambda) = \Psi_\lambda(A_\delta)$$

$$= \Psi_\lambda(q)$$

$$= \Psi_\lambda \left(\prod (e(\alpha) - e(-\alpha)) \right)$$

$$= \prod_{\alpha \in \Phi^+} \left(e^{(\lambda, \alpha)t/2} - e^{-(\lambda, \alpha)t/2} \right)$$

$$= \left(\prod (\lambda, \alpha) \right) t^{\#\Phi^+} + \text{terms of higher degree in } t.$$

Hence

$$\Psi_\delta \left(\frac{A_{\lambda+\delta}}{A_\delta} \right) = \frac{\prod (\lambda + \delta, \alpha)}{\prod (\delta, \alpha)} + \text{terms of positive degree in } t.$$

Now consider the composite homomorphism: First apply Ψ_δ and then set $t = 0$. This has the effect of replacing every $e(\mu)$ by the constant 1. Hence applied to the left-hand side of the Weyl character formula this gives the dimension of the representation $V(\lambda)$. The previous equation shows that when this composite homomorphism is applied to the right-hand side of the Weyl character formula we get the right-hand side of the Weyl dimension formula. Hence we have derived the Weyl dimension formula from the Weyl character formula.

Appendix B: Equivariant Cohomology

B.1 Hidden Symmetries

In this appendix we present some of the basic facts of equivariant cohomology from the differentiable point of view. The most important application of this topic to the subject matter of this book, the derivation of the Duistermaat–Heckman formula from Berline–Vergne localization, has been presented in Appendix 3.B. However, it appears that equivariant cohomological methods will play an increasing role in symplectic geometry and in quantization, so we thought it worthwhile to include a succinct introduction to the subject.

Most of this material is available elsewhere, for example, in the excellent book by Berline, Getzler, and Vergne [BGV] and the article [DV]. It is still instructive and inspiring to read the original 1950 work of H. Cartan [Car]. We have also relied on Kalkman's thesis [Kal] and are grateful to him for explanations of some of the material presented here.

In mechanics. Let H be a G-invariant Hamiltonian, where $G \times M \to M$ is a Hamiltonian action with moment map $\Phi \colon M \to \mathfrak{g}^*$. Assume 0 is a regular value and G compact. Then H descends to a Hamiltonian H_{red} on the reduced space

$$M_{\text{red}} = \Phi^{-1}(0)/G, \qquad \text{with } \dim M_{\text{red}} = \dim M - 2 \dim G.$$

Since M_{red} is lower dimensional, solving the equations for H_{red} involves fewer variables. In the classical literature this was regarded as a simplification. But the key point of [KKS] is that H_{red} might look much more complicated than H. For example, the Calogero system of n particles on the line with inverse square potential was proved to be completely integrable in the quantum version by Calogero, and then in the classical version by Moser. It was shown in [KKS] that the Calogero–Moser Hamiltonian was obtained from the simple quadratic

Hamiltonian involving only the kinetic energy $\frac{1}{2}\operatorname{tr}A^2$ on $\mathfrak{g}+\mathfrak{g}^*$, $\mathfrak{g}=\mathfrak{u}(n)$, when reduction was performed at a minimal orbit (rank one elements). So a straight line motion on $T^*\mathfrak{g}^*$ reduces to a complicated motion on the reduced space. The $\mathfrak{u}(n)$ symmetry is *hidden*. Similarly, although the $O(4)$ symmetry of the hydrogen atom was discovered by Pauli in 1926, it was recently discovered by [I] and [Ml] that it can be reduced from geodesic flow on a curved five-dimensional Lorentzian manifold. This is explained in [GS10], pp. 40–5.

In Quantization.   Mladenov actually uses this description of Kepler motion to give a beautiful computation of the hydrogen spectrum. It involves the "quantization of constraints." See [GS10] for an exposition. One hopes for a commutative diagram between quantization and reduction.

$$
\begin{array}{ccc}
C(M) & \xrightarrow{\text{Quant}} & Q(M) \\
R\downarrow & & \downarrow R \\
C(M)_{\text{red}} & \xrightarrow{\text{Quant}} & Q(M)_{\text{red}}.
\end{array}
$$

Indeed, this is one of the main themes of the present book. But sometimes this diagram doesn't commute or the left-hand path is not available. Then one uses the right-hand path. But quantum reduction involves some modifications, leading to BRST [KS]. This uses supersymmetry.

Birational Equivalence.   If we reduce with respect to a circle, so long as the image of the moment map is regular, the topological type of the reduced space does not change. It was shown in [GS9] that passing through a singular value corresponds to blowing up or blowing down in algebraic geometry.

Hidden Cobordism.   More generally, we can consider a new kind of cobordism, where G acts on M, and locally freely on ∂M. So we can let $Y=\partial M/G$. If this splits into two components, Y_1 and Y_2 say, this gives a "generalized cobordism" between Y_1 and Y_2. Once again, we might start with Y_1 and Y_2, and not know of G at all, so we might talk of a *hidden cobordism*. We might expect some formulas relating properties of Y to the singularities of the G action on M. First results in this direction are due to [Weit]. Later results due to [Kal] will be described in this appendix.

B.2 Equivariant Cohomology

In order to deal with the hidden cobordism, we want to invoke some homological properties of M/G when the action of G on M is not very nice. Just as in the quantization setting, we try a substitution. The idea is to replace M by $M \times E$ where E is a topological space with a free action of G and such that E is contractible. One would then replace M by $M \times E$, which has a nice quotient, and which does not change matters when the original action is free, for example. So then we define the equivariant cohomology of M by

$$H_G^*(M) := H^*((M \times E)/G).$$

How to construct E? Clearly a construction for G will also work for any subgroup of G, so to construct E for any compact group G it is enough to construct E for $U(n)$, for all n, since we can always find a faithful unitary representation of G.

Construction for S^1. Take $E = E_{S^1}$ to be the unit sphere in a complex separable Hilbert space $\mathcal{H}$. Clearly S^1 acts freely (by multiplication) on E. We must show that E is contractible. We can regard $\mathcal{H}$ as the subspace of $L^2(\mathbb{R})$ consisting of functions that vanish on the negative real axis. Choose $e \in \mathcal{H}$ such that e is supported in $[0, 1]$ and $\|e\| = 1$. We will construct a retraction of E onto e in two stages. First move $f \in E$ into

$$f_t := f(\cdot - 2t), \quad 0 \le t \le \frac{1}{2},$$

so at $t = 1/2$ its support is in $[1, \infty)$. In particular it is orthogonal to e. Then rotate

$$f_s = \cos \pi \left(s - \frac{1}{2} \right) f_{1/2} + \sin \pi \left(s - \frac{1}{2} \right) e, \quad \frac{1}{2} \le s \le 1.$$

The quotient space is the infinite-dimensional complex projective space. So if we could compute its cohomology as the limit of that of finite-dimensional projective spaces, we would expect that

$$H_G^*(\text{pt}) = \mathbb{C}[x], \quad G = S^1,$$

where x is a generator of degree 2. We will justify this conclusion in terms of an algebraic model a bit later on.

For $U(n)$ we would let E be the space of orthonormal sequences of length n in $\mathcal{H}$, where $\mathcal{H}$ is as before. Again the action is free, and the same proof, with e now an orthonormal sequence of length n supported in $[0, 1]$, proves the contractibility. This suggests that

$$H_G^{2i}(\text{pt}) = S^i(\mathfrak{g}^*)^G, \quad H_G^{2i+1}(\text{pt}) = 0.$$

The problem with the foregoing computations is that the space E is not nice. Topologists like CW complexes, for example. A lot of effort was devoted in the fifties to overcome this objection. But the resolution had already been found by H. Cartan in two papers in 1950 [Car], where the space is replaced by an algebraic model.

B.3 Superalgebras

We assume the following basic definitions (cf. [Sch] or [CNS]): For example, a commutative associative superalgebra satisfies $A = A_0 \oplus A_1$ with

$$xy = (-1)^{xy} yx.$$

We use the sloppy but short notation $(-1)^{xy}$ for $(-1)^{\deg(x)\deg(y)}$, where x and y are homogeneous elements. Similarly, a Lie superalgebra satisfies

$$[x, y] = -(-1)^{xy}[y, x] \quad \text{(superantisymmetry)}$$

and

$$[x, [y, z]] = [[x, y], z] + (-1)^{xy}[y, [x, z]] \quad \text{(superJacobi)}.$$

Here is a useful result which illustrates the simplifications introduced by superstuff:

Proposition B.3.1 *Let A be an associative superalgebra. Then* $\mathrm{Der}(A)$ *is a Lie superalgebra.*

The proof is immediate by direct verification from the definitions. We illustrate for the case of two odd derivations d_1 and d_2. We have

$$d_1 d_2(uv) = d_1[(d_2u)v + (-1)^u u d_2 v]$$
$$= (d_1 d_2 u)v + (-1)^{u+1} d_2 u d_1 v + (-1)^u d_1 u d_2 v + u d_1 d_2 v.$$

Interchanging d_1 and d_2 and adding gives

$$[d_1, d_2](uv) = (d_1 d_2 + d_2 d_1)(uv)$$
$$= ([d_1, d_2]u)v + u[d_1, d_2]v.$$

In particular, the square of an odd derivation is an even derivation. So if d is an odd derivation, to verify that $d^2 = 0$ it is enough to check this on generators.

Example (The Koszul resolution) The algebra is $\wedge \otimes S$, where $\wedge$ is the exterior algebra of a vector space and S is its symmetric algebra. The elements of S are all even, while the degrees of the elements of $\wedge$ are their exterior degrees mod 2. The elements $x \otimes 1 \in \wedge^1 \otimes S^0$ and $1 \otimes x \in \wedge^0 \otimes S^1$ generate $\wedge \otimes S$. The Koszul operator d_K is defined as the derivation extending the operator on generators given by

$$d_K(x \otimes 1) = 1 \otimes x, \quad d_K(1 \otimes x) = 0.$$

Clearly $d_K^2 = 0$ on generators, and hence everywhere.

We can also use the proposition to prove that the Koszul resolution is acyclic: Indeed, let Q be the derivation defined on generators by

$$Q(x \otimes 1) = 0, \quad Q(1 \otimes x) = x \otimes 1.$$

So $Q^2 = 0$ and $[Q, d_K] = \text{id}$ on generators. But since $[Q, d_K]$ is an even derivation, we conclude that

$$[Q, d_K] = (k + l)\,\text{id} \quad \text{on } \wedge^k \otimes S^l.$$

Thus the only cohomology of d_K lies in $\wedge^0 \otimes S^0$.

(L, G)-modules. Suppose that $L = L_0 \oplus L_1$ is a Lie superalgebra and that G is a Lie group whose Lie algebra is L_0. We also assume that we are given a representation of G on L_1 that exponentiates the (adjoint) action of L_0 on L_1 and such that the map $L_1 \otimes L_1 \to L_0$ given by the Lie bracket is a G-morphism. Such an object will be called an (L, G) pair. If $M = M_0 \oplus M_1$ is a module for L and for G such that the various consistency conditions are satisfied, then we talk of an (L, G)-module. In particular, if A is a superalgebra, we require that L act as derivations, and G act as automorphisms, and that the actions be consistent. In particular, we demand that the action of L_0 be the infinitesimal of the action of G and that the map $L \otimes A \to A$ be a G-morphism.

B.4 Differential G Complexes

Let G be a compact, connected Lie group and $\mathfrak{g}$ its Lie algebra. Consider the $\mathbb{Z}$-graded superalgebra

$$\tilde{\mathfrak{g}} = \mathfrak{g}_{-1} \oplus \mathfrak{g}_0 \oplus \mathfrak{g}_1,$$

where $\mathfrak{g}_0$ and $\mathfrak{g}_{-1}$ are copies of $\mathfrak{g}$ as vector spaces, where the bracket of two elements of $\mathfrak{g}_0$ is their bracket in $\mathfrak{g}$ and where $\mathfrak{g}_0$ acts on $\mathfrak{g}_{-1}$ via the adjoint representation. If $X \in \mathfrak{g}$ we will denote the corresponding element of $\mathfrak{g}_0$ by L_X and the corresponding element of $\mathfrak{g}_{-1}$ by ι_X. The space $\mathfrak{g}_1$ is one-dimensional with distinguished generator d, and $\mathfrak{g}_0$ acts trivially on $\mathfrak{g}_1$. Notice that since $\mathfrak{g}_2 = 0$ we have

$$d^2 = \frac{1}{2}[d, d] = 0.$$

The only brackets remaining to be specified are those between $\mathfrak{g}_1$ and $\mathfrak{g}_{-1}$ and these are defined to be

$$[d, \iota_X] = L_X. \tag{B.1}$$

A *differential G complex* (DGC) is then defined to be a commutative superalgebra with a $(\tilde{\mathfrak{g}}, G)$ action (where $\tilde{\mathfrak{g}}$ acts as derivations and G acts as automorphisms).

If the group G acts on a manifold M, then $\Omega(M)$, the algebra of differential forms on M, is a $(\tilde{\mathfrak{g}}, G)$-module: Here L_X is Lie derivative and ι_X is interior product with respect to the vector field corresponding to X, and d is exterior derivative. Of course (B.1) is just the Weil identity.

The idea of Cartan is to replace $\Omega(M)$ by more general $(\tilde{\mathfrak{g}}, G)$-modules. Then we could replace the desired space E by a suitable algebraic object. For this we need to algebraicize the notions of locally free and contractible. We will replace the notion of contractible by simply demanding that the DGC algebra A be acyclic as a d-module.

We now turn to the notion of locally free. In what follows, we will occasionally find it convenient to write things in terms of a basis. So we let $X_1, \ldots, X_n$ be a basis of the Lie algebra $\mathfrak{g}$ and c^i_{jk} be the structure constants relative to this basis, so

$$[X_j, X_k] = c^i_{jk} X_i,$$

where the Einstein summation convention is in force wherever there are Latin superscripts and subscripts.

Recall that the action of G on a manifold is called *locally free* if the evaluation map of the Lie algebra $\mathfrak{g}$ to the tangent space at every point is an injection. This implies that we can find linear differential forms θ^a such that

$$\iota_a \theta^b = \delta^a_b. \tag{B.2}$$

(Here and in what follows we will write

$$\iota_a \text{ for } \iota_{X_a} \text{ and } L_a \text{ for } L_{X_a}.)$$

We take (B.2) as the definition of locally free for a general DGC A. That is, A is called *locally free* if there are elements $\theta^a \in A$ such that (B.2) holds. As G is compact, by averaging over the group if necessary, we can arrange that the θs transform like the coadjoint representation. We then have

$$\begin{aligned}
-c^a_{bk} \theta^k &= L_b \theta^a \\
&= \iota_b d\theta^a + d\iota_b \theta^a \\
&= \iota_b d\theta^a.
\end{aligned}$$

Therefore,

$$d\theta^a = -\frac{1}{2}c^a_{jk}\theta^j\theta^k + \mu^a, \quad \text{where } \iota_b\mu^a = 0, \ \forall a, b. \tag{B.3}$$

Here the μ^a are even (sometimes called the curvature forms) and, since d commutes with the group action, must also transform like the coadjoint representation. It follows from (B.3) and $d^2 = 0$ that $d\mu^a$ is some linear combination of $\mu^b\theta^j$. To see what this combination is, we use the fact that the μ transform like the coadjoint representation:

$$\begin{aligned}
-c^a_{bj}\mu^j &= L_b\mu^a \\
&= \iota_b d\mu^a + d\iota_b\mu^a \\
&= \iota_b d\mu^a.
\end{aligned}$$

So

$$d\mu^a = -c^a_{jk}\theta^j\mu^k. \tag{B.4}$$

So the plan is to find a DGC A that is acyclic, then replace $\Omega(M)$ by $\Omega(M) \otimes A$ and "pass to the quotient." For this we need an algebraic substitute for passing to the quotient. If N is a manifold on which G acts freely with $\pi: N \to Y := N/G$ the projection onto the quotient, then $\pi^*: \Omega(Y) \to \Omega(N)$ is an injection. Its image consists of all forms ω that satisfy

$$\iota_X\omega = 0, \quad L_X\omega = 0, \quad \forall X \in \mathfrak{g}. \tag{B.5}$$

Condition (B.5) makes sense for any DGC, and if ω satisfies (B.5) so does $d\omega$. An ω satisfying (B.5) will be called *basic*, and the algebra of all basic elements of A will be denoted by A_{basic}. So the plan is to find an acyclic DGC B and then define

$$H^*_G(M) := H^*((\Omega(M) \otimes B)_{\text{basic}}).$$

In fact, we will construct B to be bigraded over the integers and will show that its cohomology is concentrated in bidegree $(0, 0)$. More generally, if A is any DGC, we define

$$H^*_G(A) := H^*((A \otimes B)_{\text{basic}}).$$

We will show, using the Matthai–Quillen isomorphism (see Section B.6), that if A is locally free, then

$$H^*((A \otimes B)_{\text{basic}}) = H^*(A_{\text{basic}}).$$

This will then prove that our definition is independent of the choice of the model B.

B.5 The Weil Algebra

The Weil algebra is

$$W(\mathfrak{g}) := \wedge(\mathfrak{g}^*) \otimes S(\mathfrak{g}^*)$$

with $\mathfrak{g}_{-1}$ acting via interior product on the $\wedge(\mathfrak{g}^*)$ component.

$$\iota_X(\Theta \otimes f) = (\iota(X))\Theta) \otimes f.$$

The algebra $\mathfrak{g}_0$ acts via the extension (by derivation) of the coadjoint representation. The operator d is given by

$$d_W = d_{CE} + d_K,$$

where d_{CE} is the usual Chevalley–Eilenberg Lie algebra differential (cf. (B.9)) and where d_K (the Koszul operator) is the derivation extending the linear map

$$\wedge^1 \otimes S^0 \to \wedge^0 \otimes S^1,$$
$$x \otimes 1 \mapsto 1 \otimes x, \quad 1 \otimes x \mapsto 0$$

that we defined earlier.

In terms of a basis $X_1 \ldots, X_n$ of $\mathfrak{g}$ with corresponding structure constants c^i_{jk}, let $\theta^1, \ldots, \theta^n$ be the dual basis of $\wedge^1(\mathfrak{g}^*) \otimes S^0$ and $\mu^1, \ldots, \mu^n$ the dual basis in $\wedge^0 \otimes S^1(\mathfrak{g}^*)$ so that the θ^a and μ^b generate W as an algebra. As usual, we write ι_a instead of ι_{X_a} and L_a instead of L_{X_a}. Then on the generators we have (using the Einstein summation convention and dropping wedges and tensor product signs)

$$d\theta^a = -\frac{1}{2}c^a_{jk}\theta^j\theta^k + \mu^a$$

$$d\mu^a = -c^a_{jk}\theta^j\mu^k, \text{ as we have argued,}$$

so

$$\iota_b d\theta^a = -\frac{1}{2}c^a_{bk}\theta^k + \frac{1}{2}c^a_{jb}\theta^j$$

$$= -c^a_{bk}\theta^k$$

$$= L_a\theta^b$$

and

$$\iota_b d\mu^a = -c^a_{bk}\mu^k$$

$$= L_b\mu^a.$$

This proves that

$$[\iota_X, d] = L_X$$

on generators and hence on all elements. The fact that $d^2 = 0$ follows from Jacobi's identity. A good way to see all this is to identify $W(\mathfrak{g})$ as the equivariant cohomology of the group G acting on itself by left multiplication. (See Section B.9 on the equivariant cohomology of homogeneous spaces.)

Since

$$\iota_a \theta^b = \delta_a^b$$

the Weil algebra is locally free. We can prove that it is acyclic by the usual Weil argument for double complexes, starting from the fact that d_K is acyclic: Suppose that ω is of total degree $k + 2l$ so

$$\omega = \omega_{k,l} + \omega_{k+2,l-1} + \cdots .$$

Then

$$d_W \omega = 0 \implies d_K \omega_{k,l} = 0.$$

If $(k, l) \neq (0, 0)$ this implies that there is a $\mu_{k+1,l-1}$ with

$$d_K \mu_{k+1,l-1} = \omega_{k,l}.$$

Then

$$\omega - d_W \mu_{k+1,l-1} = \omega_{k+2,l-1} - d_{CE} \mu_{k+1,l-1} + \cdots$$

has a lower l degree, and proceeding inductively shows that ω is a coboundary.

For any differential G complex B we define its *equivariant cohomology* as

$$H_G^*(B) = H^*[(W \otimes B)_{\text{basic}}]. \tag{B.6}$$

We now can compute $H_G^*(\text{pt}) = H^*(W_{\text{basic}})$. Indeed, the elements of W that are annihilated by all the ι_X constitute $1 \otimes S(\mathfrak{g}^*)$. So the invariant elements constitute $S(\mathfrak{g}^*)^G$ (we are assuming that G is connected). On them, the operator d is trivial. So

$$H_G^*(\text{pt}) = S(\mathfrak{g}^*)^G.$$

For the case of a semisimple Lie group this is the same as

$$S(t^*)^W,$$

where T is a maximal torus and W is the Weyl group.

For a general B, the computation again involves two steps: the computation of first the "horizontal elements" of $W \otimes B$, the elements annihilated by ι, and then, among them, the invariant elements. So the first step is to find the intersection of the kernels of

$$\iota_X \otimes 1 + 1 \otimes \iota_X \quad \forall X \in \mathfrak{g}.$$

This intersection is hard to compute. Instead, we shall find an automorphism of $W \otimes B$ that carries

$$\iota_X \otimes 1 + 1 \otimes \iota_X \mapsto \iota_X \otimes 1, \quad \forall X \in \mathfrak{g}.$$

B.6 The Matthai–Quillen Isomorphism

Let A and B be DGCs and suppose that

$$\theta^a \in A^{\text{odd}}, \iota_b \theta^a = \delta^a_b.$$

Then

$$\theta^\alpha \otimes \iota_\alpha \quad \text{(no summation)}$$

defines an even derivation of $A \otimes B$:

$$\begin{aligned}
\theta \otimes \iota[(a_1 \otimes b_1)(a_2 \otimes b_2)] &= \theta \otimes \iota[(-1)^{b_1 a_2}(a_1 a_2 \otimes b_1 b_2)] \\
&= (-1)^{a_1 + a_2 + b_1 a_2} \theta a_1 a_2 \\
&\quad \otimes [(\iota b_1) b_2 + (-1)^{b_1} b_1 \iota b_2] \\
&= [(-1)^{a_1} \theta a_1 \otimes \iota b_1][a_2 \otimes b_2] \\
&\quad + [a_1 \otimes b_1][(-1)^{a_2} \theta a_2 \otimes \iota b_2] \\
&= [(\theta \otimes \iota)(a_1 \otimes b_1)][a_2 \otimes b_2] \\
&\quad + [a_1 \otimes b_1][(\theta \otimes \iota)(a_2 \otimes b_2)].
\end{aligned}$$

We have

$$(\theta^\alpha \otimes \iota_\alpha) \cdot (\theta^\beta \otimes \iota_\beta) = -\theta^\alpha \theta^\beta \otimes \iota_\alpha \iota_\beta$$

so

$$(\theta^\alpha \otimes \iota_\alpha) \cdot (\theta^\beta \otimes \iota_\beta) = (\theta^\beta \otimes \iota_\beta) \cdot (\theta^\alpha \otimes \iota_\alpha)$$

and

$$(\theta^\alpha \otimes \iota_\alpha)^2 = 0.$$

Hence

$$\phi \overset{\text{def}}{=} \exp(\theta^a \otimes \iota_a) = \prod_\alpha (1 + \theta^\alpha \otimes \iota_\alpha) \in \text{Aut}(A \otimes \Omega(B))$$

is well defined and

$$\phi^{-1} = \exp(-\theta^a \otimes \iota_a) = \prod (1 - \theta^\alpha \otimes \iota_\alpha).$$

Proposition B.6.1 *Suppose that B is generated by elements that satisfy*

$$\iota_a f = 0 \ \text{or} \ \iota_a \iota_b \omega = 0 \ \forall a, b.$$

Then

$$\begin{array}{ccc}
A \otimes B & \xrightarrow{\phi} & A \otimes B \\
\iota_a \otimes 1 + 1 \otimes \iota_a \ \downarrow & & \downarrow \ \iota_a \otimes 1 \\
A \otimes B & \xrightarrow{\phi} & A \otimes B
\end{array} \tag{B.7}$$

for all a.

Proof. The derivation

$$\gamma = \theta^a \otimes \iota_a \tag{B.8}$$

(summation convention in force) vanishes on all elements of the form $w \otimes f$. For elements of the form $1 \otimes \omega$, we have

$$\gamma(1 \otimes \omega) = \theta^a \otimes \iota_a \omega,$$
$$\gamma^2(1 \otimes \omega) = 0,$$

so

$$\phi(1 \otimes f) = 1 \otimes f \ \text{if} \ \iota_a f = 0 \ \forall a,$$
$$\phi(1 \otimes \omega) = 1 \otimes \omega + \theta^a \otimes \iota_a \omega, \ \text{if} \ \iota_a \iota_b \omega = 0 \ \forall a, b.$$

These elements generate $A \otimes B$. So it is enough to check (B.7) on them. For the first type it is obvious. For elements of the form $1 \otimes \omega$ we have

$$\begin{aligned}
(\iota_a \otimes 1)\phi(\omega) &= (\iota_a \otimes 1)(1 \otimes \omega + \theta^b \otimes \iota_b \omega) \\
&= 1 \otimes \iota_a \omega \\
&= (\iota_a \otimes 1 + 1 \otimes \iota_a)(1 \otimes \omega) \\
&= \phi[(\iota_a \otimes 1 + 1 \otimes \iota_a)(1 \otimes \omega)]. \quad \square
\end{aligned}$$

In other words, ϕ carries the horizontal elements of $A \otimes B$ into the elements that satisfy $(\iota_a \otimes 1)\sigma = 0$. So, for example,

$$\phi(W \otimes B) = S(\mathfrak{g}^*) \otimes B.$$

Also, since ϕ is intrinsically defined, it is a morphism of $(\tilde{\mathfrak{g}}, G)$-modules. Hence

$$\phi((W \otimes B)_{\text{basic}}) = (S(\mathfrak{g}^*) \otimes B)^G.$$

We must compute the conjugate of the differential under ϕ, that is, we must compute

$$\phi(d_W \otimes 1 + 1 \otimes d_M)\phi^{-1}.$$

We claim that

$$\phi d\phi^{-1} = d + \theta^a \otimes L_a - \mu^a \otimes \iota_a. \tag{B.9}$$

Proof. Recalling that $d = d_A \otimes 1 + 1 \otimes d_B$ on $A \otimes B$ we have (no summation convention)

$$[\theta^\alpha \otimes \iota_\alpha, d] = -d\theta^\alpha \otimes \iota_\alpha + \theta^\alpha \otimes L_\alpha.$$

Also,

$$[\theta^b \otimes \iota_b, d\theta^\alpha \otimes \iota_\alpha] = \theta^b d\theta^\alpha \otimes [\iota_b, \iota_\alpha] = 0,$$

while

$$\begin{aligned}
[\theta^\beta \otimes \iota_\beta, \theta^\alpha \otimes L_\alpha] &= -\theta^\alpha \theta^\beta \otimes [L_\alpha, \iota_\beta] \\
&= -\sum_a c^a_{\alpha\beta} \theta^\alpha \theta^\beta \otimes \iota_a
\end{aligned}$$

and

$$\begin{aligned}
[\theta^c \otimes \iota_c, \theta^a \theta^b \otimes \iota_f] &= \theta^c \theta^a \theta^b \otimes [\iota_c, \iota_f] \\
&= 0.
\end{aligned}$$

It follows that $(\mathrm{ad}\,\gamma)^3 d = 0$ and hence

$$\begin{aligned}
\phi d\phi^{-1} &= \exp(\mathrm{ad}\,\gamma)d \\
&= d + [\gamma, d] + \frac{1}{2}[\gamma, [\gamma, d]] \\
&= d - \mu^a \otimes \iota_a + \theta^a \otimes L_a. \qquad \square
\end{aligned}$$

B.7 The Cartan Model

In the case $A = W$, the horizontal elements of $W \otimes B$ are transformed by ϕ into elements of $S(\mathfrak{g}^*) \otimes B$. On $S(\mathfrak{g}^*)$ we have

$$d_W = d_{CE} = \theta^a L_a.$$

Hence on horizontal elements we have

$$\begin{aligned}
\phi d\phi^{-1} &= \theta^a L_a \otimes 1 + \theta^a \otimes L_a - \mu^a \otimes \iota_a + 1 \otimes d_B \\
&= (\theta^a \otimes 1)(L_a \otimes 1 + 1 \otimes L_a) + 1 \otimes d_a - \mu^a \otimes \iota_a.
\end{aligned}$$

The image of the basic complex consists of the invariant elements of $S(\mathfrak{g}^*) \otimes B$ on which the first term in the previous expression vanishes. Hence we define d_C, the Cartan operator, to be the restriction of $\phi d \phi^{-1}$ to the image of the basic complex. We have

$$d_C = 1 \otimes d_B - \mu^a \otimes \iota_a, \quad [S(\mathfrak{g}^*) \otimes B]^G \to [S(\mathfrak{g}^*) \otimes B]^G. \qquad (\text{B}.10)$$

This is the *Cartan model* for equivariant cohomology.

B.8 Locally Free Complexes

As a second application of the Matthai–Quillen isomorphism, let us compute the equivariant cohomology for the locally free case and show that it is the basic cohomology:

Proposition B.8.1 *If A is ($\mathbb{Z}$-graded and) locally free then $H_G^*(A) = H^*(A_{\text{basic}})$.*

Proof. This time we take $B = W$ and define ϕ as before. Then ϕ carries $(A \otimes W)_{\text{basic}}$ onto

$$(A_{\text{hor}} \otimes W)^G$$

and d into (B.9). Let

$$C^{i,j} = (A_{\text{hor}}^i \otimes W^j)^G$$

and write the operator (B.9) as

$$d + \theta^a \otimes L_a - \mu^a \otimes \iota_a = \delta' + \delta'',$$

where

$$\delta' = d_M \otimes 1 + \theta^a \otimes L_a - \mu^a \otimes \iota_a \colon C^{ij} \to C^{i+1,j} \oplus C^{i+2,j-1}$$

and

$$\delta'' = 1 \otimes d_W \colon C^{i,j} \to C^{i,j+1}.$$

Since the Weil complex is acyclic, we conclude by the usual staircase argument that the cohomology of $\phi d \phi^{-1}$ is the homology of the complex concentrated along the bottom row. Since L_a and ι_a vanish on $W^0 = S^0(\mathfrak{g}^*) \otimes \wedge^0(\mathfrak{g}^*)$ the proposition follows. $\qquad \square$

B.9 Equivariant Cohomology: Homogeneous Spaces

In this section we work directly with the definition (B.9). We follow closely the treatment in [DV]. So let M be a manifold with an action of connected Lie group, G:

$$G \times M \to M.$$

On $C^\infty(\mathfrak{g}, \Omega(M))$ define d_C by (B.9), which can be expressed in a basis-free form as

$$d_C(\alpha)(X) = d(\alpha(X)) - \iota(X_M)\alpha(X),$$

where X_M denotes the vector field on M corresponding to $X \in \mathfrak{g}$. So

$$(d_C^2\alpha)(X) = -L_{X_M}\alpha(X), \qquad (B.11)$$

where $L_\bullet$ denotes the Lie derivative.

Let $\Omega_G(M)$ denote the space of G equivariant elements of $C^\infty(\mathfrak{g}, \Omega(M))$. Then for any $Y \in g$ and $\alpha \in \Omega_G(M)$ we have

$$L_{Y_M}\alpha(X) = \alpha([Y, X]).$$

In particular,

$$L_{X_M}\alpha(X) = 0$$

and so

$$d_C^2 = 0 \quad \text{on } \Omega_G(M).$$

As we saw in the preceding section, the corresponding cohomology groups are the equivariant cohomology groups of M. On the subcomplex of polynomial maps we have a $\mathbb{Z}$ gradation (by using twice the polynomial gradation as before, and the usual exterior gradation). With respect to this gradation d_C has degree $+1$.

Functorialities

If $H \to G$ is a group homomorphism with $\mathfrak{h} \to \mathfrak{g}$ the corresponding homomorphism of Lie algebras, then composition induces a map

$$C^\infty(\mathfrak{g}, \Omega(M)) \to C^\infty(\mathfrak{h}, \Omega(M))$$

and hence induces a map of the complexes

$$\Omega_G(M) \to \Omega_H(M),$$

which induces a morphism

$$H_G^*(M) \to H_H^*(M).$$

Also if $N \to M$ is a morphism of G-manifolds, then pullback induces a morphism

$$H_G^*(M) \to H_G^*(N).$$

Homogeneous spaces

Let us describe the Cartan complex of a homogeneous space. In the process we will find that the Cartan complex for G acting on itself by left multiplication is just the Weil complex $W(\mathfrak{g})$.

So let G be a Lie group and H a closed subgroup, with Lie algebras $\mathfrak{g}$ and $\mathfrak{h}$. Set $M = G/H$ and $e \in M$ the point $\{H\}$. We may identify

$$TM_e = \mathfrak{g}/\mathfrak{h},$$

so the cotangent space at e is identified with $(\mathfrak{g}/\mathfrak{h})^*$. Let

$$\rho: \mathfrak{g} \to \mathfrak{g}/\mathfrak{h}$$

denote the projection. For any $X \in \mathfrak{g}$, the infinitesimal generator of the one parameter group $\exp(-tX)$ of transformations on M is denoted by X_M. So the value of the vector field X_M at the point e is given by

$$X_M(e) = -\rho(X).$$

Let α be a G-equivariant (smooth or polynomial) map of $\mathfrak{g}$ into $\Omega(M)$, and let $\alpha_{[p]}$ denote the component in $\Omega^p(M)$. Then for $a \in G$ and $\xi_1, \ldots, \xi_p \in TM_e$ the equivariance disentangles into

$$(\alpha_{[p]}(\mathrm{Ad}(a)X))_{a \cdot e}(da(\xi_1), \ldots, da(\xi_p)) = \alpha_{[p]}(X)_e(\xi_1, \ldots, \xi_p). \quad \text{(B.12)}$$

So let

$$\alpha \mapsto \tilde{\alpha}$$

denote evaluation at e, and

$$T_H = (S(\mathfrak{g}^*) \otimes \wedge(\mathfrak{g}^*))^H.$$

Then $\alpha \mapsto \tilde{\alpha}$ gives an isomorphism of T_H with the space of polynomial maps of $\mathfrak{g}$ into forms on M that are G-equivariant. We wish to see what the operator d_C goes into under this identification.

Let ι be the derivation on the space of equivariant forms given by

$$(\iota\alpha)(X) = \iota(X_M)\alpha(X)$$

(where the expression on the right denotes interior product). So the corresponding derivation of T_H is given by

$$\tilde{\iota}\tilde{\alpha}(X) = -\iota(\rho(X))\tilde{\alpha}(X).$$

Suppose we pick a basis $E_1, \ldots, E_s$ of G whose first r elements form a basis of $\mathfrak{h}$. Let x^i denote the corresponding coordinates, so the x^i, $i > r$, form a basis of $(\mathfrak{g}/\mathfrak{h})^* = \mathfrak{h}^0 \subset \mathfrak{g}^*$. Let

$$d_K = \sum_{j>r} x^j \iota(\rho E_j).$$

It is a derivation with square zero of $S(\mathfrak{g}^*) \otimes \wedge (\mathfrak{g}/\mathfrak{h})^*$ and $\tilde{\iota}$ is the restriction of $-\iota$ to the subspace T_H.

Proposition B.9.1 *Under the identification of $\Omega_G(G/H)$ with T_H, the operator d_C becomes identified with the operator*

$$d_{CE} + d_K,$$

where d_{CE} is the restriction of the usual Chevalley–Eilenberg operator computing the cohomology of $\mathfrak{g}$ with values in the $\mathfrak{g}$-module. In particular, taking H to be the trivial subgroup, we see that the equivariant complex of G acting on itself by left multiplication is just the Weil complex W.

We have already seen that the derivation $-\iota$ goes over into d_K. So we must check that the usual d operator goes over into the Chevalley–Eilenberg operator whose definition we now recall: Let V be a $\mathfrak{g}$-module, and define the operator d_{CE} on $V \otimes \wedge (\mathfrak{g}^*)$,

$$d_{CE}\colon \wedge^p (\mathfrak{g}^*) \otimes V \rightarrow \wedge^{p+1}(\mathfrak{g}^*) \otimes V,$$

by

$$d_{CE}\alpha(Y_1, \ldots, Y_{p+1}) = \sum_i (-1)^{i+1} Y_i \cdot \alpha(Y_1, \ldots, \hat{Y}_i, \ldots, Y_{p+1})$$
$$+ \sum_{i<j} \alpha([Y_i, Y_j], Y_1, \ldots, \hat{Y}_i,$$
$$\ldots, \hat{Y}_j, \ldots, Y_{p+1}). \qquad (B.13)$$

Notice that if $\alpha \in \wedge((\mathfrak{g}/\mathfrak{h})^*) \otimes V$, which means that $\alpha(Y_1, \ldots, Y_p) = 0$ if any of the $Y_i \in \mathfrak{h}$, then $d_{CE}\alpha$ has the same property. This is what is meant by the phrase "restriction" in the proposition.

By pullback from the projections $G \rightarrow G/H$ and $\mathfrak{g} \rightarrow \mathfrak{g}/\mathfrak{h}$ it is enough to prove the proposition for the case when H is the trivial subgroup. Let $Y_1, \ldots, Y_{p+1}$ be elements of $\mathfrak{g}$, also thought of as left invariant vector fields on G. Let X be an element of $\mathfrak{g}$. Then (B.12) gives

$$\alpha(X)_a(Y_2, \ldots, Y_{p+1}) = \tilde{\alpha}(\mathrm{Ad}_{a^{-1}} X)(Y_2, \ldots, Y_{p+1}), \quad \forall a \in G.$$

Taking
$$a = \exp t Y_1$$
in this formula, differentiating with respect to t, and setting $t = 0$ gives

$$Y_1 \cdot \big(\alpha(X)(Y_2, \dots, Y_{p+1})\big)$$
$$= \frac{d}{dt}\tilde{\alpha}(\mathrm{Ad}_{\exp -tY_1}X)(Y_2, \dots, Y_{p+1})\big|_{t=0}. \qquad (B.14)$$

The expression on the left of (B.14) is the Lie derivative of the function

$$\alpha(X)(Y_2, \dots, Y_{p+1})$$

with respect to the vector field Y_1. It is one of the terms that occurs in the formula for the exterior derivative of $\alpha(X)$:

$$d(\alpha(X))(Y_1, \dots, Y_{p+1})$$
$$= \sum_i (-1)^{i+1} Y_i \cdot \alpha(X)(Y_1, \dots, \hat{Y}_i, \dots, Y_{p+1})$$
$$+ \sum_{i<j} (-1)^{i+j} \alpha(X)([Y_i, Y_j], Y_1, \dots, \hat{Y}_i, \dots, \hat{Y}_j, \dots, Y_{p+1}). \qquad (B.15)$$

The expression on the right of (B.14) is the action of Y_1 on the element

$$\tilde{\alpha}(\bullet)(Y_2, \dots, Y_{p+1}) \in S(\mathfrak{g}^*)$$

that occurs in (B.9) with $V = S(\mathfrak{g}^*)$. So comparing (B.15) with (B.9) using (B.14) proves the proposition.

As a corollary we obtain another proof of $d_W^2 = 0$ in the Weil complex.

Let us now turn to the case of a homogeneous space G/H, where H is compact. According to the proposition, we have identified the equivariant complex with

$$W_H \overset{\text{def}}{=} (S(\mathfrak{g}^*) \otimes \wedge(\mathfrak{h}^0))^H$$

equipped with the Chevalley–Eilenberg differential. Suppose we choose an H-invariant embedding $\mathfrak{h}^* \to \mathfrak{g}^*$ onto a complement of $\mathfrak{h}^0$. This allows us to identify

$$S(\mathfrak{g}^*) \sim S(\mathfrak{h}^*) \otimes S(\mathfrak{h}^0).$$

Then the restriction of the Koszul operator from $S(\mathfrak{g}^*) \otimes \wedge(\mathfrak{g}^*)$ to $S(\mathfrak{g}^*) \otimes \wedge(\mathfrak{h}^0)$ is just the extension of the Koszul operator of the vector space $\mathfrak{h}^0$, acting on

$$S(\mathfrak{h}^0) \otimes \wedge(\mathfrak{h}^0)$$

to $S(\mathfrak{g}^*) \otimes \wedge(\mathfrak{h}^0)$, where we regard $S(\mathfrak{g}^*)$ as an $S(\mathfrak{h}^*)$ module. So from the acyclicity of the Koszul operator of $\mathfrak{h}^0$ we conclude that the d_K cohomology of W_H vanishes except in bidegree $(2i, 0)$ and that

$$H^{2i,0}(W_H, d_K) = S^i(\mathfrak{h}^*)^H.$$

The usual double complex argument then gives the equivariant cohomology of a homogeneous space as

$$H_G^{2i}(G/H) = S^i(\mathfrak{h}^*)^H, \qquad H_G^{2i+1}(G/H) = 0. \tag{B.16}$$

It is interesting to compare the proof that we gave earlier for the computation of the equivariant cohomology of locally free actions with a more "geometric" proof using the computations of this section.

Suppose that we have a general G action, $G \times M \to M$, and let $p \in M$ and $X = G \cdot p$ the orbit through p. We can then (via the exponential map for a G-invariant Riemann metric) find a G-invariant neighborhood U of X with a G-equivariant retract of U onto X. Hence

$$H_G^*(U) = S(\mathfrak{h}^*)^H,$$

where

$$H = G_p$$

is the isotropy group of p. In particular, if the action is locally free so $\mathfrak{h} = 0$,

$$H_G^0(U) = \mathbb{C}, \qquad H_G^i(M) = 0, \quad i > 0.$$

We can conclude from the usual Mayer–Vietoris argument that

$$H_G^*(M) = H^*(M/G)$$

if the action is (globally) free, so $Y = M/G$ is a manifold. Indeed, if

$$\omega \in \Omega^*(Y)$$

then, if $\pi: M \to Y$ denotes the canonical projection,

$$\pi^* \in S^0(\mathfrak{g}^*) \otimes \Omega^*(M)$$

is annihilated by all L_a and ι_a so π^* induces a homomorphism

$$\pi^*: H^*(Y) \to H_G^*(M).$$

By the preceding remark this is an isomorphism on contractible open sets of Y. So if we assume (for simplicity) that we can cover Y by finitely many such open sets with all intersections contractible, then the five lemma together with the Mayer–Vietoris sequence of the base Y and the G-equivariant version up on M proves that π^* is an isomorphism.

For locally free actions, this mode of argument does not immediately work, since even the existence of a simple cover is not obvious. However, the algebraic argument given earlier works, and somehow replaces the Mayer–Vietoris style argument.

B.10 The Thom Form According to Mattai–Quillen

A fundamental construction in differential topology is the Thom class of a vector bundle (cf. [BT]). Let $P \to X$ be a principle $K = SO(n)$ bundle and

$$U = (P \times V)/K, \quad U \to X$$

the associated vector bundle, where V is the standard representation space of $SO(n)$. We can identify forms on U with K-equivariant basic forms on $P \times V$ (in the Cartan model) and integration over U corresponds to integration over V.

In general, there is no canonical way of constructing a representative differential form in $\Omega(U)$ for the Thom class. However, Matthai and Quillen have shown how to construct a canonical representative form in $(S(\mathfrak{k}^*) \otimes \Omega(P \times V)))^K$ for the Thom class. We have

$$(S(\mathfrak{k}^*) \otimes \Omega(V))^K \subset (S(\mathfrak{k}^*) \otimes \Omega(P \times V))^K$$

as a subcomplex, and hence to find a Thom form on U it suffices to find a

$$d_C = d - x^a \otimes \iota_a$$

closed form τ in $(S(\mathfrak{k}^*) \otimes \Omega(V)^K$ with nonvanishing fiber integral.

Let $v_1, \ldots, v_n$ be an oriented orthonormal basis of V, let $u_1, \ldots, u_n$ be the corresponding coordinates, and let $\psi^1, \ldots, \psi^n$ be the dual basis of V^*. The *Berezin integral*

$$\int \mathcal{D}\psi \colon \wedge V^* \to \mathbb{R}$$

picks out the coefficient of $\psi^1 \wedge \cdots \wedge \psi^n$, and so depends on the metric and orientation. It extends to a superalgebra homomorphism:

$$\int \mathcal{D}\psi \colon \wedge V^* \otimes A \to A$$

for any commutative superalgebra A. The Mattai–Quillen form is defined as

$$\tau = e^{-\frac{1}{4}\sum u_i^2} \int \mathcal{D}\psi \left[\exp\left(\psi^i du_i + \sum_i \psi^i x^a L_a \psi^i \right) \right],$$

where the L_a range over a basis of k and the x^a over the dual basis. The highest-order expression in the wedges of the dus comes from ignoring the terms in the exponential involving the xs and is hence

$$e^{-\frac{1}{4}\sum u_i^2} du_1 \wedge \cdots \wedge du_n.$$

This is a convergent integral with nonzero value. So all we must do is check that τ is d_C closed. It is therefore enough to prove that

$$d_C\left(e^{-\frac{1}{4}\sum u_i^2}\left[\exp\left(\psi^i du_i + \sum_i \psi^i x^a L_a \psi^i\right)\right]\right)$$

is a sum of terms of the form

$$\left(\frac{\partial}{\partial \psi^i}\right)\alpha_i,$$

where $(\partial/\partial \psi^i)$ denotes interior product with respect to the basis element v_i in $\wedge V^*$, since applying such an operator kills the top order term, and hence applying Berezin integration will yield zero.

So let

$$\sigma := \left[\exp\left(\psi^i du_i + \sum_i \psi^i x^a L_a \psi^i\right)\right].$$

Clearly

$$d\sigma = 0,$$

and hence

$$d\left[e^{-\frac{1}{4}\sum u_i^2}\sigma\right] = \left[-\frac{1}{2}\sum u_i du_i\right] \wedge e^{-\frac{1}{4}\sum u_i^2}\sigma.$$

Applying the derivation $x^\beta \iota_\beta$ passes right through the Gaussian term. For any derivation s and any even nilpotent element η in any superalgebra we have

$$s \exp \eta = (s\eta) \exp \eta.$$

The interior product ι_β only hits the du_i in the exponential and

$$\iota_\beta du_i = L_\beta u_i.$$

Hence

$$x^b \iota_b \exp\left[\psi^i du_i + \sum_i \psi^i x^a L_a \psi^i\right]$$

$$= (-\psi^i x^b L_b u_i) \exp\left[\psi^i du_i + \sum_i \psi^i x^a L_a \psi^i\right].$$

But since each of the L_b belongs to $\mathfrak{so}(n)$, we have

$$\psi^i L_b u_i = -u_i L_b \psi.$$

Hence

$$d_{\mathbb{C}}\left[e^{-\frac{1}{4}\sum u_i^2}\sigma\right]$$

$$= -e^{-\frac{1}{4}\sum u_i^2}\sum u_i\left(\frac{1}{2}du_i + x^a L_a \psi^i\right)\exp\left[\psi^i du_i + \sum_i \psi^i x^a L_a \psi^i\right]$$

$$-\frac{1}{2}\left[e^{-\frac{1}{4}\sum u_i^2}\sigma\right]\sum v_i\left(\frac{\partial}{\partial\psi^i}\right)\exp\left[\psi^i du_i + \sum_i \psi^i x^a L_a \psi^i\right],$$

proving that τ is $d_{\mathbb{C}}$ closed.

B.11 Equivariant Superconnections

Let $E \to M$ be a G-superbundle, so $E = E_+ \oplus E_-$ is a $\mathbb{Z}_2$ graded vector bundle on which G acts as even morphisms. Let $\Omega(M, E)$ denote the space of E-valued differential forms on M. So $\Omega(M, E)$ is an $\Omega(M)$-module and is spanned by expressions of the form $\alpha \otimes s$, where $\alpha \in \Omega(M)$ and s is a section of E. If s is a section of E_+ and α is a differential form of degree k, then $\alpha \otimes s$ is given degree $k \bmod 2$. If s is a section of E_- then α is given degree $k+1 \bmod 2$. This gives a $\mathbb{Z}_2$-grading to $\Omega(M, E)$, and makes $\Omega(M, E)$ into a supermodule over $\Omega(M)$.

We can also consider the bundle $\mathrm{End}(E)$ of superalgebras, and the superalgebra $\Omega(M, \mathrm{End}(E))$. Then $\Omega(M, E)$ is a module over $\Omega(M, \mathrm{End}(E))$. The elements of $\Omega(M, \mathrm{End}(E))$ can be characterized inside $\mathrm{End}[\Omega(M, E)]$ as those operators that supercommute with $\Omega(M)$. (Indeed, they are characterized as the operators commuting with multiplication by functions.)

An odd operator $A \in \mathrm{End}[\Omega(M, E)]$,

$$A\colon \Omega^+(M, E) \to \Omega^-(M, E), \quad A\colon \Omega^-(M, E) \to \Omega^+(M, E),$$

is called a *superconnection* if

$$A(\alpha \wedge \sigma) = d\alpha \wedge \sigma + (-1)^{\deg(\alpha)}\alpha \wedge A\sigma,$$

$$\forall \alpha \in \Omega(M), \sigma \in \Omega(M, E). \tag{B.17}$$

We can write this as

$$[A, \alpha] = d\alpha \tag{B.18}$$

if we think of α as an element on $\mathrm{End}[\Omega(M, E)]$ given by left multiplication.

Let ξ be a vector field on M. Then the interior product by $\iota\xi$, denoted by $i(\xi)$, acts as a derivation on $\Omega(M)$ and also as an odd operator on $\Omega(M, E)$ if we define it by

$$\iota(\xi)(\alpha \otimes s) = \iota(\xi)\alpha \otimes s.$$

It then follows that as elements of $\text{End}[\Omega(M, E)]$ we have

$$[\iota(\xi), a] = \iota(\xi)\alpha.$$

Thus we have

$$[A, [\iota(\xi), \alpha]] = [[A, \iota(\xi)], \alpha] - [\iota(\xi), [A, \alpha]]$$

or

$$[[A, \iota(\xi)], \alpha] = d\iota(\xi)\alpha + \iota(\xi)d\alpha = L_\xi\alpha$$

by Weil's identity.

Since G acts as morphisms of E, it acts as even morphisms of $\Omega(M, E)$. We will call the corresponding action of $\mathfrak{g}$ the "Lie derivative," and denote the action of $X \in \mathfrak{g}$ by L_X. Similarly, we will denote the Lie derivative of forms on M with respect to the vector field X_M by L_X. With this notation, we have

$$[L_X, \alpha] = L_X\alpha$$

as elements of $\text{End}[\Omega(M, E)]$.

Thus we see from the preceding equation that

$$[[A, \iota(X_M)] - L_X, \alpha] = 0 \ \forall \alpha \in \Omega(M).$$

Let us define

$$\mu(X) = L_X - [A, \iota X]. \tag{B.19}$$

Since $\mu(X)$ commutes with all of $\Omega(M)$ and is an even operator,

$$\mu(X) \in \Omega^+(M, \text{End}(E)).$$

Now A^2 is also an even operator, and

$$\begin{aligned}
[A^2, f] &= A[A, f] + [A, f]A \\
&= [A, df] \\
&= d^2 f \\
&= 0,
\end{aligned}$$

for any function f. Hence

$$A^2 \in \Omega(M, \text{End}(E)).$$

Notice that

$$A^2 + \mu(X) = (A - \iota_X)^2 + L_X. \tag{B.20}$$

Hence

$$\begin{aligned}
[A - \iota_X, A^2 + \mu(X)] &= [a - \iota_X, (A - \iota_X)^2 + L_X] \\
&= [A, L_X].
\end{aligned}$$

So if A is G-equivariant (as we shall assume from now on) we get the Bianchi identity

$$[A - \iota_X, A^2 + \mu(x)] = 0. \tag{B.21}$$

B.12   Equivariant Characteristic Classes

Let C be an associative superalgebra and $B = B^+ \oplus B^-$ a commutative superalgebra. By a *state* of C with values in B we mean a linear function

$$l \colon C \to B \qquad \text{such that} \qquad l([a, b]) = 0 \ \ \forall a, b \text{ in } C.$$

For example, if $V = V^+ \oplus V^-$ is a supervector space and $C = \mathrm{End}(V)$, then the *supertrace* mapping $C \to \mathbb{R}$ given by

$$\mathrm{Tr} \begin{pmatrix} A & B \\ C & D \end{pmatrix} = \mathrm{tr}\, A - \mathrm{tr}\, D$$

in the obvious block notation is a state from C to the superalgebra $\mathbb{R}$ (with vanishing odd part). Tensoring with differential forms gives the supertrace

$$\mathrm{Tr} \colon \Omega(M, \mathrm{End}(E)) \to \Omega(M).$$

Let A be an equivariant superconnection and p any polynomial. Then

$$p(A^2 + \mu(\cdot)) \in S[\mathfrak{g}^*] \otimes \Omega(M, \mathrm{End}(E)).$$

Taking the supertrace, we get

$$\mathrm{Tr}\, p(A^2 + \mu(\cdot)) \in S[\mathfrak{g}^*] \otimes \Omega(M). \tag{B.22}$$

We claim that

$$d_C\, \mathrm{Tr}\, p(A^2 + \mu(\cdot)) = 0 \tag{B.23}$$

and that the cohomology class of $\mathrm{Tr}\, p(A^2 + \mu(\cdot))$ is independent of the choice of equivariant superconnection. The proofs depend on the following two lemmas:

Lemma B.12.1 *For any $\sigma \in \Omega(M, \mathrm{End}(E))$ we have*

$$d\, \mathrm{Tr}\, \sigma = \mathrm{Tr}[A, \sigma].$$

Proof. Locally we have

$$A = d + \omega, \quad \omega \in \Omega(M, \text{End}(E)),$$

and the superbundle E is trivialized. In terms of this trivialization we can think of σ as a supermatrix-valued differential form, in which case

$$[d, \sigma] = d\sigma.$$

Hence

$$\text{Tr}[A, \sigma] = \text{Tr}\, d\sigma + \text{Tr}[\omega, \sigma].$$

The second term vanishes and $d\,\text{Tr}\,\sigma = \text{Tr}\,d\sigma$, proving the lemma. $\square$

Now ι_X acts only on the differential form component and hence commutes with supertrace:

$$\iota_X \text{Tr}\,\sigma = \text{Tr}\,\iota_X\sigma = \text{Tr}\,[\iota_X, \sigma].$$

Thus

$$d_C \text{Tr}\, p(A^2 + \mu(\cdot))(X) = \text{Tr}\,[A - \iota_X, \text{Tr}\, p(A^2 + \mu(X))].$$

We must prove that this vanishes. For this it is enough to prove that it vanishes when $p(z) = z^n$ is a polynomial. But this is a consequence of Bianchi's identity (B.21).

Lemma B.12.2 *Let σ_t be a one-parameter family of elements of $\Omega(M, \text{End}(E))$. Then*

$$\frac{d}{dt} \text{Tr}\, p(\sigma_t) = \text{Tr}\left(\frac{d}{dt}\sigma_t\right) p'(\sigma_t).$$

Proof. Check this on monomials where it is obvious. $\square$

Now let A_t be a one-parameter family of equivariant superconnections, so

$$\mu_t(X) = L_X - [A_t, \iota_X].$$

Then

$$\frac{d}{dt}(A_t^2 + \mu_t(X)) = \left[A_t - \iota_X, \frac{d}{dt}A_t\right].$$

From this we get, using the lemmas,

$$\frac{d}{dt}\operatorname{Tr} p\big(A_t^2 + \mu_t(\cdot)\big) = \operatorname{Tr}\left[A_T - \iota_X, \left(\frac{dA_t}{dt}\right)\right]p'\big(A_t^2 + \mu_t(\cdot)\big)$$

$$= d_C \operatorname{Tr}\left(\frac{dA_t}{dt}\right)p'\big(A_t^2 + \mu_t(\cdot)\big).$$

Hence

$$\operatorname{Tr} p\big(A_1^2 + \mu_1(\cdot)\big) - \operatorname{Tr} p\big(A_0^2 + \mu_0(\cdot)\big) = d_C \int \operatorname{Tr}\left(\frac{dA_t}{dt}\right)p'\big(A_t^2 + \mu_t(\cdot)\big)dt.$$

Since we can join any two superconnections by a line segment of superconnections $A_t = tA_1 + (1-t)A_0$ this proves that the cohomology class is independent
of the choice of superconnections.

Chern characters

We can consider entire functions instead of polynomials p and so define the
Chern character as

$$\operatorname{ch}(E, A) = \operatorname{Tr} e^{-(A^2 + \mu(\cdot))}.$$

If E and F are superbundles with connections A and B then clearly

$$\operatorname{ch}(E \oplus F, A \oplus B) = \operatorname{ch}(E, A) + \operatorname{ch}(F, B).$$

On the other hand $A \otimes 1 + 1 \otimes B$ is a superconnection on $E \otimes F$ and

$$(A \otimes 1 + 1 \otimes B)^2 = A^2 \otimes 1 + 1 \otimes B^2$$

from which it follows that the Chern character is multiplicative on tensor products.

B.13   Reduction Formula for Locally Free Torus Actions

We begin with $G = S^1$ acting locally freely on M. Let v be a basis of $\mathfrak{g} = \mathbb{R}$,
and x the dual basis of $\mathfrak{g}^*$. We choose a "curvature form" θ that is G-invariant
with

$$\iota_v\theta \equiv 1.$$

So

$$d_C\theta = d\theta - x.$$

The element $d_C\theta$ is not invertible in the Cartan complex,

$$\tilde{\Omega} = \Omega(M)^G \otimes \mathbb{C}[x]$$

but becomes invertible in the localized ring $\tilde{\Omega}_x$ since we have

$$\frac{1}{d_C\theta} = \frac{1}{d\theta - x}$$

$$= -\frac{1}{x}\left(\frac{1}{1 - d\theta/x}\right)$$

$$= -\frac{1}{x}\left(1 + \frac{d\theta}{x} + \left[\frac{d\theta}{x}\right]^2 + \cdots\right).$$

Thus if $\omega \in \tilde{\Omega}_x$ satisfies

$$d_C\omega = 0,$$

then

$$\omega = d_C\gamma, \quad \gamma = \frac{\theta\omega}{d_C\theta}. \tag{B.24}$$

Suppose that we start with $\omega \in \tilde{\Omega}$. Then the left-hand side of (B.24) has no negative powers of x. So if we let γ_{hol} denote the sum of the terms involving nonnegative powers of x in the expansion of γ and γ_{res} the coefficient of x^{-1} we have

$$\omega = d_C\gamma_{\text{hol}} - \iota_v\gamma_{\text{res}}. \tag{B.25}$$

Remarks

1. Suppose that $\omega \in \tilde{\Omega}^p$, where $p \geq \dim M$. Expanding $\theta\omega$ out in powers of x shows that $\gamma = \gamma_{\text{hol}}$ so $\beta = 0$ in this case.
2. Multiplication by θ followed by the interior product ι_v amounts to taking the negative of the horizontal component. Also,

$$\iota_v d\theta = \iota_v d\theta + d\iota_v\theta$$

$$= L_v\theta$$

$$= 0.$$

Hence (in the case of general m) we can write

$$\beta = \left(\frac{\omega_{\text{hor}}}{d_C\theta}\right)_{\text{res}}.$$

Alternatively we can write this as

$$\beta = \pi^*\frac{1}{2\pi}\pi_*\left(\frac{\theta\omega}{d_C\theta}\right)_{\text{res}}.$$

3. Let K be some compact, connected Lie group and $G = S^1$. Assume that both G and K act on M and that these actions commute. Suppose that the action of G is locally free as before. We continue the preceding notation of v a basis of $\mathfrak{g}$, x, and θ. Let $w_1, \ldots, w_n$ be a basis of $\mathfrak{k}$, $\iota_a = \iota_{w_a}$, and $y^1, \ldots, y^n$ the dual basis of $\mathfrak{k}^* = S^1(\mathfrak{k}^*)$ and define

$$\phi_a = \iota_a \theta, \quad \mu = d\theta - y^a \phi_a.$$

Now

$$\tilde{\Omega} = \left(\Omega(M) \otimes \mathbb{C}[x] \otimes S(\mathfrak{k}^*)\right)^{G \times K}$$

and

$$d_C = d - x\iota_v - y^a \iota_a.$$

As before, if $d_C \omega = 0$ we can write (B.24) in the formal power series completion of $\tilde{\Omega}_X$, where γ now has the expansion

$$\gamma = -\frac{\theta \omega}{x}\left(1 + \frac{\mu}{x} + \frac{\mu^2}{x^2} + \cdots\right).$$

If the $\mathfrak{k}$ action is such that

$$\iota_a \theta = 0, \tag{B.26}$$

so the $\phi_a = 0$, there is no need to pass to a formal power series completion.

4. As an application of the preceding remark, suppose we have a torus action that is locally free, and we choose connection forms $\theta^1, \ldots, \theta^n$ with

$$\iota_a \theta^b = \delta_a^b,$$

where $v_1, \ldots, v_n$ is a basis of the Lie algebra, ι_a the corresponding interior products, and $x^1, \ldots, x^n$ the dual basis in $S[\mathfrak{g}^*]$. Then we can apply the foregoing procedure inductively to obtain that any equivariantly closed form ω differs by a closed form from

$$\beta = \left(\frac{\omega_{\text{hor}}}{(x^1 - d\theta^1) \cdots (x^n - d\theta^n)}\right)_{\text{res}},$$

where now res means the coefficient of $(x^1 \cdots x^n)^{-1}$. For example, any polynomial $p \in \mathbb{C}[x^1, \ldots, x^n]$ can be thought of as an equivariantly closed form (with no $\Omega(M)$ component of positive degree and the constant function on M). For example, suppose that $p(x) = x^l$ is a monomial. Then the preceding formula gives

$$\beta = d\theta^l.$$

In other words, the Chern–Weil homomorphism itself appears as this computation in equivariant cohomology!

5. If we let Q denote the operation of left multiplication by $\theta/d_C\theta$ then instead
    of (B.24) we can write the homotopy equation

$$\omega = d\,Q\omega + Qd\omega,$$

which is valid for all ω.

B.14 Localization

We continue to let $G = S^1$ and M be a compact, oriented manifold on which
G acts. Let M^G denote the fixed-point set of this action, and

$$M_0 = M/M^G.$$

Since G is one-dimensional, the action of G on M_0 is locally free. If we assume
that M^G is a union of proper submanifolds, then the integral of any form ω over
M is the same as its integral over M_0. If

$$\omega \in \tilde{\Omega}^l, \quad l \geq \dim M,$$

then we can apply (B.24) to conclude that

$$\int_M \omega = -\lim_{\epsilon \to 0} \int_{\partial M_\epsilon} \gamma,$$

where M_ϵ is a small neighborhood of M^G and ∂M_ϵ its boundary. The Berline–
Vergne localization formula will compute this limit in terms of integrals over
the connected components of M^G. The formula asserts that

$$\int_M \omega = \sum_X \int_X \frac{i^*\omega}{e(X)}, \tag{B.27}$$

where the sum is over all connected components of M^G and $e(X)$ denotes the
equivariant Euler class of X. The purpose of this section is to give a computational proof of this formula under a mild splitting assumption on the components.

Let X be a connected component of M^G. By the exponential map for a
G-invariant metric, we can identify a tubular neighborhood of X with an open
neighborhood of the zero section in the normal bundle $E \to X$. For each $x \in X$
the fiber E_x of the normal bundle carries a representation of G which can have
no fixed vectors. (Otherwise the image of the line of fixed vectors under the exponential map would be a curve of fixed vectors in M.) So E_x decomposes into a
direct sum of two-dimensional subspaces. Let us make the simplifying assumption that the representations of G on each of these subspaces are nonequivalent.
By a choice of orientation on each subspace, this gives a decomposition of E_x

into a direct sum of complex line bundles, which, by continuity, gives a global decomposition of E. So we make the weaker simplifying assumption that we have the global decomposition

$$E = L_1 \oplus \cdots \oplus L_p. \tag{B.28}$$

Let m_i denote the weight of the representation of G on L_i, and let θ_i denote a G-invariant connection form on the principal bundle P_i, where P_i is the unit circle bundle in L_i, on which we have chosen a G-invariant metric $\| \cdot \|_i$. Let

$$s_i = \frac{1}{2} \| \cdot \|_i^2,$$

thought of (by projection onto the ith factor) as a function on E, and let

$$s = s_1 + \cdots + s_p.$$

We claim that

$$\theta = \frac{1}{s} \left(\frac{s_1}{m_1} \theta_1 + \cdots + \frac{s_p}{m_p} \theta_p \right) \tag{B.29}$$

is a connection form defined on all of $E/0$. It suffices to check this locally, where each $P_i = U \times S^1$ and

$$\theta_i = d\phi_i + \pi^* \alpha_i,$$

where ϕ_i is the angular variable and the second term is the pullback of a form on U. In coordinates on the complex line,

$$d\phi = (x\, dy - y\, dx)/r^2,$$

so multiplying by $s = r^2/2$ gives the form$(x\, dy - y\, dx)/2$, which is smooth across the origin. Hence the form (B.29) is smooth outside the zero section, and since it is G invariant and

$$v = \sum_i m_i \frac{\partial}{\partial \phi_i}$$

it is a connection form. We will take these connection forms (one for each component of M^G) and extend them (by a G-invariant partition of unity, say) to get a connection form on all of M_0. So we now have

$$\int_M = -\lim_{\epsilon \to 0} \sum_X \int_{\partial X_\epsilon} \gamma,$$

where X_ϵ denotes the ϵ-tubular neighborhood of the zero section of the normal bundle, relative to the metric we have chosen, and we sum over all components of the fixed-point set. We compute each summand as follows: Write

$$-\gamma = \frac{\theta \wedge \omega}{x - d\theta}$$

and

$$\omega = \pi^* i^* \omega + (\omega - \pi^* i^* \omega).$$

The forms θ and $x - d\theta$ are invariant under homotheties, and the form $(\omega - \pi^* i^* \omega)$ tends to zero as $\epsilon \to 0$. Hence we are reduced to considering the integral

$$- \int_{\partial X_\epsilon} \gamma \pi^* i^* \omega,$$

and since the integrand is invariant under homothety, we may take $\epsilon = 1$. Now the vector field v vanishes along X, and hence $d_C = d$ on X. Thus the condition $d_C \omega = 0$ implies that $d i^* \omega = 0$. So if we expand $i^* \omega$ in powers of x, each coefficient is a closed form, and we can consider each term separately. So we may assume that

$$i^* \omega = \omega_j x^r, \quad \omega_j \in \Omega^j(X), \quad d\omega_j = 0,$$

and our assumption on degrees says that $j + 2r \geq \dim M$. The integrand involves only the term that has the same degree as

$$\dim \partial X_1 = \dim M - 1.$$

So the integral we must compute is

$$x^{r-k-1} \int_{\partial X_1} \theta \wedge (d\theta)^p \wedge \pi^* \omega_j, \quad j + 2k + 2 = \dim M. \qquad (B.30)$$

Since $j + 2r \geq \dim M$, the condition $j + 2k + 2 = \dim M$ implies that $r \geq k+1$, so the exponent of x occurring in (B.30) is nonnegative. On ∂X_1 we have $s = 1$ so

$$\theta = \sum_{i=1}^{p} \frac{s_i}{m_i} \theta_i,$$

$$d\theta = \sum_{i=1}^{p} \frac{ds_i}{m_i} \theta_i + \sum_{i=1}^{p} \frac{s_i}{m_i} d\theta_i,$$

and we can expand $(d\theta)^k$ using the binomial formula for the sum of the two terms occurring on the right in the last equation. Since each $d\theta_i$ is horizontal, the only nontrivial contribution to the integral in (B.30) must involve the $(p-1)$st power of $\sum_{i=1}^{p} \frac{ds_i}{m_i}\theta_i$ in the binomial expansion in order to get $2p-1$ "vertical" degrees. Hence the integrand in (B.30) is

$$\binom{k}{p-1}\left(\sum_{i=1}^{p}\frac{s_i}{m_i}\theta_i\right)\left(\sum_{i=1}^{p}\frac{ds_i}{m_i}\theta_i\right)^{p-1}\left(\sum_{i=1}^{p}\frac{s_i}{m_i}d\theta_i\right)^{k-p+1}\pi^*\omega_j. \quad (B.31)$$

Notice that the last two factors in this expression are horizontal, and their degrees add up to $2k-2p+2+j = \dim M - 2p = \dim X$. If we expand the fourth factor using the multinomial formula, it becomes

$$\sum_{|I|=k-p+1}\frac{(k-p+1)!}{I!}\frac{s^I}{m^I}(d\theta)^I.$$

The coefficient of $(d\theta)^I \wedge \omega_j$ in this expression, multiplied by the first three factors in (B.31), contributes to the integral only by restricting to the fiber and integrating over the fiber, as $(d\theta)^I \wedge \omega_j$ uses up all the horizontal degrees. The first three terms in (B.31) multiply out to

$$\frac{1}{m_1\cdots m_p}\binom{k}{p-1}(p-1)!\theta_1\wedge\cdots\wedge\theta_k\sum(-1)^{i-1}s_i ds_1$$

$$\wedge\cdots\wedge\widehat{ds_i}\wedge\cdots\wedge ds_p.$$

Upon restriction to the fiber, the θ_i become the angular variables, and hence the integral of $\theta_1\wedge\cdots\wedge\theta_p$ contributes $(2\pi)^p$ to the fiber integral. Hence the integral (B.30) becomes

$$(2\pi)^p\binom{k}{p-1}(p-1)!\frac{1}{m_1\cdots m_p}(k-p+1)!\sum_{|I|=k-p+1}\frac{c_I}{I!m^I}\int_X(d\theta)^I\omega_j,$$

$$(B.32)$$

where

$$c_I = \int_\Delta s^I\sum(-1)^i s_i ds_1\wedge\cdots\wedge\widehat{ds_i}\wedge\cdots ds_p$$

and Δ is the $(p-1)$-simplex in $\mathbb{R}^p$ given by

$$\Delta = \{(s_1,\ldots,s_p)\mid s_i\geq 0, s_1+\cdots+s_p = 1\}.$$

Let us compute this integral. Let

$$w = \sum s_i\frac{\partial}{\partial s_i}$$

be the Euler vector field giving infinitesimal dilatation, so that

$$L_w(s^I ds_1\wedge\cdots\wedge ds_p) = (|I|+p)s^I ds_1\wedge\cdots\wedge ds_p.$$

On the other hand,

$$L_w(s^I ds_1 \wedge \cdots \wedge ds_p) = d\iota_w(s^I ds_1 \wedge \cdots \wedge ds_p)$$
$$= d[s^I \sum (-1)^i s_i ds_1 \wedge \cdots \wedge \widehat{ds_i} \wedge \cdots ds_p]$$

and hence

$$c_I = (|I| + p) \int_{\mathbf{S}_p} s^I ds_1 \cdots ds_p,$$

where $\mathbf{S}_p$ is the solid simplex

$$\mathbf{S}_p = \{s_i \geq 0 \mid s_1 + \cdots + s_p \leq 1\}.$$

We claim, by induction on p, that

$$\int_{\mathbf{S}_p} s^I ds_1 \cdots ds_p = \frac{I!}{(|I| + p)!}.$$

Indeed, this is clearly true for $p = 1$. Assume that it is true for p, and let $J = (j, I)$ be a multi-index in $p + 1$ dimensions. Then,

$$\int_{\mathbf{S}_{p+1}} s^J ds_1 \cdots ds_{p+1} = \int_0^1 s^j \left(\int_{(1-s)\mathbf{S}_p} s^I ds_1 \cdots ds_p \right) ds$$

$$= \frac{I!}{(|I| + p)!} \int_0^1 s^j (1 - s)^{|I| + p} ds$$

$$= \frac{I!}{(|I| + p)!} \times \frac{j!(|I| + p)!}{(|I| + p + j + 1)!}$$
(evaluating the Beta integral)

$$= \frac{J!}{(|J| + p + 1)!}.$$

Substituting

$$c_I = \frac{I!}{(|I| + p - 1)!} = \frac{I!}{k!}$$

into (B.32) yields a cancellation of all the factorials and the coefficient of x^{r-k-1} is

$$\frac{(2\pi)^p}{m_1 \cdots m_p} \int_X \omega_j \wedge \sum \frac{(d\theta)^I}{m^I}.$$

Now $|I| = k - P + 1$ so we can write $-k - 1$ as $-p - |I|$ and hence the integral over ∂X_ϵ can be written as

$$\frac{(2\pi)^p}{x^p m_1 \cdots m_p} \int_X \sum \frac{(d\theta)^I}{m^I x^{|I|}} i^* \omega. \tag{B.33}$$

Notice that this formula now works independently of the j (the degree of the homogeneity in the exterior degree of $i^*\omega$) when we use the usual convention that the integration over X picks out only the terms of correct degree. In other words, the formula (B.33) now works for general ω.

Under our assumption that the normal bundle splits into a sum of line bundles, the equivariant Euler class is (up to a factor of $(-1)^p$) the product of the Euler classes of each of the line bundles, and each of these is given by

$$\frac{1}{2\pi} d_C \theta_i = d\theta_i - m_i x.$$

Hence

$$(-1)^p \frac{1}{e(X)} = \frac{(2\pi)^p}{\prod(m_i x - d\theta_i)}$$

$$= \frac{(2\pi)^p}{m_1 \cdots m_p x^p} \prod \left(\frac{1}{1 - \frac{d\theta_i}{m_i x}} \right)$$

$$= \frac{(2\pi)^p}{m_1 \cdots m_p x^p} \sum \frac{(d\theta)^I}{m^I x^{|I|}}.$$

Substituting this into (B.33) allows us to rewrite (B.33) as

$$\int_X \frac{i^*\omega}{e(X)}. \tag{B.34}$$

This completes the proof of (B.27).

Now suppose that M is a manifold with boundary ∂M and $M \cap M^G = \emptyset$ so that G acts locally freely on the boundary. Then the proof given before of the Berline–Vergne localization formula shows that

$$\int_M \omega + \int_{\partial M} \frac{\theta \wedge \omega}{x - d\theta} = \sum_X \int_X \frac{i^*\omega}{e(X)}. \tag{B.35}$$

In this equation, both sides should be thought of as elements of $\mathbb{C}[x, x^{-1}]$.

Since G acts locally freely on ∂M, the space $Y = \partial M / G$ is an orbifold, and we have an identification of $H_G^*(\partial M)$ with $H^*(Y)$. At the level of forms, this identification is given by the residue formula of the preceding section. Let

$$r_* : \tilde{\Omega}(M) \to \Omega(Y)$$

be the composition of the restriction map $\tilde{\Omega}(M) \to \tilde{\Omega}(\partial M)$ with the identification via the residue formula so that

$$r(\omega) = \pi^* \frac{1}{2\pi} \pi_* i^* \left(\frac{\theta\omega}{d_C\theta} \right)_{\text{res}}.$$

Taking residues of both sides of (B.35) gives Kalkman's formula [Kal]

$$\int_Y r(\omega) = \sum_X \left(\int_X \frac{i^*\omega}{e(X)} \right)_{\text{res}}. \tag{B.36}$$

B.15 Localization in Stages

We begin by giving the version of (B.35) that applies when there is a compact, connected group K that acts on M and whose action commutes with the action of $G = S^1$. We continue the assumptions about the action of G on M and ∂M. We also retain the notation of Remark 3 of the section before last. Thus $\omega = d\gamma$, where

$$\gamma = \frac{\theta\omega}{d_{\mathbb{C}}\theta} = -\frac{\theta\omega}{x - \mu}, \quad \mu = d\theta - y^a \phi_a,$$

with

$$\phi_a = \iota_a \theta.$$

Then the proof of (B.35) (by Stokes's theorem as in the preceding section) shows that

$$\int_M \omega + \int_{\partial M} \frac{\theta\omega}{x - \mu} = \sum_X \int_X \frac{i^*\omega}{\bar{e}(X)}, \tag{B.37}$$

where $\bar{e}(X)$ is the $(G \times K)$-equivariant Euler class of the normal bundle to X. Similarly, we get an equivariant version of Kalkman's formula

$$\int_Y r(\omega) = \sum_X \left(\int_X \frac{i^*\omega}{\bar{e}(X)} \right)_{\text{res}}. \tag{B.38}$$

The map r going from $\tilde{\Omega}(M)_{G \times K}$ to $\tilde{\Omega}(Y)_K$ is given by restriction followed by passage to the quotient as before. Both sides of (B.38) should be regarded as elements of $\mathbb{C}[y^1, \ldots, y^n]^K$. If K contains a central circle subgroup, then we can further compute the integrals on the right by localization.

Appendix C: Update

The subject matter of this book is an active area of research, and much progress has been achieved since the latest draft has been written. As authors we are thus faced with the standard dilemma: Choose between waiting until the field stabilizes or writing a timely but incomplete book. We have opted for the second choice. As a partial palliative we indicate some recent research in this short appendix knowing full well that we are omitting important results (apologies to the various authors) and that much of what we say here will be outdated by the time this book actually appears in print.

C.1 Commutativity of Quantization and Reduction

Let G be a compact connected Lie group acting equivariantly on a line bundle $L \to M$ preserving a connection ∇ whose curvature form ω is symplectic, so that the action of G on (M, ω) is Hamiltonian with moment map Φ. Suppose that 0 is a regular value of the moment map, and let $Z = \Phi^{-1}(0)$ and $i \colon Z \to M$ be the inclusion map. The infinitesimal action of G on sections of L is given by Kostant's formula

$$D_v s := \nabla_v s + 2\pi i \langle \Phi, v \rangle s, \quad v \in \mathfrak{g}.$$

Therefore sections of $i^* L$ are G-invariant iff they are autoparallel. The orbit space $L_G = i^* L / G$ is a line bundle over the Marsden–Weinstein reduced space M_G and there is a unique connection ∇_G on the line bundle

$$L_G \to M_G$$

with the property that

$$\pi^* \nabla_G = i^* \nabla.$$

Since 0 is a regular value of the moment map, the action of G on Z and on L is locally free, so L_G and M_G are orbifolds, in general. If we assume that the action is free, these become manifolds.

Geometrical quantization is a catch-all phrase for finding a consistent method Q for associating a unitary representation of G to the geometric data (L, ∇). The commutativity mentioned in the title of this subsection is the formula

$$\dim Q(M_G) = \dim Q(M)^G. \tag{C.1}$$

The $Q(M_G)$ occurring on the left-hand side of this equation is the Hilbert space associated by the quantization procedure to the data (L_G, ∇_G, M_G). The $Q(M)^G$ occurring on the right is the space of G-invariants in the unitary representation space associated to (L, ∇, M) by Q. More generally, we may want to consider the objects on both sides as virtual representations. Of course this formula acquires content only with a choice of quantization procedure Q. Here is one such choice:

Choose an almost complex structure J on M that is compatible with its symplectic form ω and such that g defined by $g(u, v) := \omega(Ju, v)$ is positive definite. The complexified tangent bundle splits into $\pm i$ eigenbundles of J:

$$TM \otimes \mathbb{C} = T^{1,0} \oplus T^{0,1},$$

and so there is a bigrading

$$\Omega^k = \bigoplus_{p+q=k} \Omega^{p,q}$$

of the de Rham complex and we have the operators d and $\delta = d^*$ (given by the Riemannian metric g). Let d'' be the $(0, 1)$ component of d and δ'' be the $(0, -1)$ component of δ. Then

$$\delta'' + d'': \Omega^{0,\pm} \to \Omega^{0\mp},$$

where $\Omega^{0,+}$ is the even part and $\Omega^{0,-}$ is the odd part of $\Omega^{0,*}$. Coupled with the connection ∇ we get an operator

$$\partial\!\!\!/_{\mathbb{C}}: L \otimes \Omega^{0,+} \to L \otimes \Omega^{0,-}$$

called the spin$^{\mathbb{C}}$ Dirac operator. It is an elliptic operator and we define

$$Q(M) := [\ker \partial\!\!\!/_{\mathbb{C}}] - [\operatorname{coker} \partial\!\!\!/_{\mathbb{C}}] \tag{C.2}$$

as a virtual representation.

In the Kählerian setting the conjecture (C.1) was proved in [GS3]. In [GS4] a version of (C.1) was proved; namely, it was shown that (C.1) is true if the original line bundle L is replaced by a suitably high power L^n. (The definition of Q there is slightly different than the one just given.)

Recently, substantial progress has been made on this conjecture, most of it based on a localization theorem of Jeffrey, Kirwan, and Witten [JK1]. For the case of Abelian groups, a version of this conjecture was established in [Gil2] under additional hypotheses, using the results of [JK1]. The full Abelian version of the conjecture was proved independently by Meinrenken [Mein1] and Vergne [Ver]. A short, simple proof of this result using the symplectic cutting technique of Lerman [L] was obtained by Duistermaat, Guillemin, Meinrenken, and Wu [DGMW].

As to non-Abelian groups, the conjecture has been proved for rank-one groups by Jeffrey and Kirwan [JK2]. Meinrenken [Mein1] has shown that the result of Guillemin–Sternberg [GS4] on high powers of the line bundle can be derived by a fairly elementary stationary-phase argument. Martin and Weitsman [MW] were able to get a reasonable estimate for the power n: By hypothesis, the origin is a regular value of the moment map, hence contained in an open convex polytope consisting of regular values. Raising L to the power n multiplies the moment map, and hence dilates this polytope of regular values by a factor of n. They showed that if n is so large that this dilated polytope contains the sum of the positive roots, and all its images under the Weyl group, then the conjecture holds. Finally Meinrenken [Mein2] proved the conjecture in the general case.

C.2 Toric Varieties

In the special case where the group is a torus and the action is completely integrable, the image of the moment map completely determines the Hamiltonian group action according to a classical theorem of Delzant [Del]. The subject matter then becomes the theory of toric varieties, from a symplectic point of view. All multiplicities are zero or one and the Duistermaat–Heckman measure is also zero or one (times Lebesgue measure); it is just Lebesgue measure on the image polytope of the moment map and zero outside. These issues relate to algebraic geometry (the geometry of toric varieties) and number theory (the counting of lattice points in polyhedra) as well as representation theory. See, for example, [CS], [KK], [KP], [Pom], and the recent book [Gil1].

C.3 Orbifolds

One of the technical complications we have cited is that when 0 is a regular value of the moment map Φ, one can only conclude that the action of G on $Z = \Phi^{-1}(0)$ is locally free, so the reduced space $M_G = Z/G$ will be an orbifold, but not necessarily a manifold. Thus we require an extension of the theory of Hamiltonian group actions to symplectic orbifolds. Foundational results for this theory can be found in [LT]. For example, the Atiyah–Guillemin–Sternberg theorem on the convexity of the image of the moment map still holds. Delzant's theorem on symplectic toric manifolds is modified as follows: There is a one-to-one correspondence between symplectic toric orbifolds and convex rational simple polytopes with positive integers attached to each facet.

C.4 Presymplectic Hamiltonian Actions

An important generalization of the techniques of this book is achieved when the closed form ω is not assumed to be nondegenerate, but the existence of an equivariant moment map Φ for the group action, that is, a map $\Phi: M \to \mathfrak{g}^*$ satisfying $d\langle\Phi, \xi\rangle = -\iota(\xi_M)\omega$, is assumed. The Duistermaat–Heckman measure, that is, the push-forward by the moment map of $\omega^n/n!$, where $n = 1/2 \dim M$, now becomes a signed measure. For torus actions it is still piecewise polynomial and the boundaries of the regions where it is a polynomial are contained in a finite union of hyperplanes. In particular, for completely integrable torus actions one obtains twisted polytopes as the support of the D–H measure instead of convex polytopes. There is still a tight relation between the representation theory (via the Atiyah–Bott fixed-point formula) and the symplectic geometry. For all this see [KT] and [GK].

C.5 Nonisolated Fixed Points

The main formulas in Chapter 3 have been generalized in [GC] to the case where the fixed points need not be isolated. On the symplectic side, the "Heckman formula" must be modified using techniques analogous to the Minkowski theory of mixed volumes, whereas on the representation-theory side, the Atiyah–Bott fixed-point formula must be replaced by the Atiyah–Segal–Singer formula.

Bibliography

[AM]   Abraham, R., and Marsden, J. E. *Foundations of Mechanics*. Reading, MA: Benjamin/Cummings, 1978.

[AS]   Ambrose, W., and Singer, I. M. A theorem on holonomy. *Trans. AMS* **75** (1953), 428–43.

[At]   Atiyah, M. F. Convexity and commuting Hamiltonians. *Bull. London Math. Soc.* **14** (1982), 1–15.

[AB]   Atiyah, M. F., and Bott, R. The Lefschetz fixed point formula for elliptic complexes I. *Ann. Math.* **86** (1967), 374–407.

[BBB]   Berard Bergery, L., and Bourguignon, J.-P. Laplacians and Riemannian submersions with totally geodesic fibres. *Illinois Journal of Mathematics* **26** (1982), 181–200.

[BGV]   Berline, N.; Getzler, E.; and Vergne, M. *Heat Kernels and Dirac Operators*. New York: Springer-Verlag, 1992.

[BV]   Berline, N., and Vergne, M. Classes charactéristiques équivariantes. Formules de localisation en cohomologie équivariante. *C.R. Acad. Sci. Paris* **295** (1982), 539–41.

[BGG]   Bernstein, I. N.; Gelfand, I. M.; and Gelfand, S. I. Structure of representations generated by vectors of highest weight. *Func. Anal. Appl.* **5** (1971), 1–8;
Differential operators on the base affine space and a study of g-modules. In *Lie Groups and Their Representations*, I. M. Gelfand, ed., 21–64. New York: Halsted, 1975.

[Ber]   Berry, M. V. Classical adiabatic angles and quantum adiabatic phase. *J. Phys. A* **18** (1985), 15–27.

[BH]   Berry, M. V., and Hannay, J. H. Classical non-adiabatic angles. *J. Phys. A* **21** (1988), 325–31.

[Bes]   Besse, A. *Einstein Manifolds*. New York: Springer-Verlag, 1987.

[Bo1]   Bott, R. An application of Morse theory to the topology of Lie groups. *Bull. Soc. Math. France* **84** (1956), 251–82.

[Bo2]   Bott, R. The geometry and representation theory of compact Lie groups. In *Representation Theory of Lie Groups: Proceedings of the SRC/LMS Research Symposium on Representations of Lie Groups, Oxford, 28 June–15 July 1977*, M. F. Atiyah et al., eds. New York: Cambridge University Press, 1979.

[BT]   Bott, R., and Tu, L. W. *Differential Forms in Algebraic Topology*. New
            York: Springer-Verlag, 1982.

[CS]   Cappell, S. E., and Shaneson, J. L. Genera of algebraic varieties and
            counting lattice points. *Bull. A.M.S.* **30** (1994), 62–9.

[Car]   Cartan, H. La transgression dans un groupe de Lie et dans un espace
            fibré principal, in *Colloque de Topologie (Espaces Fibrés) Tenu à
            Bruxelles du 5 au 8 Juin 1950*, C.B.R.M., 57–71. Liège: Georges
            Thone, 1951.

[CNS]   Corwin, L.; Ne'eman, Y.; and Sternberg, S. Graded Lie algebras in
            mathematics and physics (Bose–Fermi symmetry). *Rev. Modern Phys.*
            47 (1975), 573–603.

[Del]   Delzant, T. Hamiltoniens périodiques et images convexes de
            l'application moment. *Bul. Soc. Math. France* **116** (1988), 315–39.

[DV]   Duflo, I. M., and Vergne, M. Cohomologie équivariant et descente.
            Astérisque **215** (1993), 5–107.

[Duis]   Duistermaat, J. J. On the existence of global action–angle variables.
            Comm. Pure Appl. Math. **33** (1980), 687–706.

[DGMW]   Duistermaat, J. J.; Guillemin, V.; Meinrenken, E.; and Wu, S.
            Symplectic reduction and Riemann–Roch for circle actions. *Math.
            Research Letters* **2** (1995), 256–66.

[DH1]   Duistermaat, J. J., and Heckman, G. On the variation of cohomology of
            the symplectic form of the reduced phase space. *Invent. Math.* **69**
            (1982), 259–68.

[DH2]   Duistermaat, J. J., and Heckman, G. Addendum to "On the variation of
            cohomology of the symplectic form of the reduced phase space."
            Invent. Math. **72** (1983), 153–8.

[FH]   Fulton, W., and Harris, J. *Representation Theory: A First Course*. New
            York: Springer-Verlag, 1991.

[Go]   Godement, R. *Topology algébrique et théorie des faisceaux*. Paris:
            Herman, 1957.

[GKM]   Golin, S.; Knauf, A.; and Marmi, S. The Hannay angles: geometry,
            adiabaticity and an example. *Comm. Math. Phys.* **123** (1989), 95–123.

[GOL]   Golubitsky, M., and Schaeffer, D. *Singularities and Groups in
            Bifurcation Theory*. New York: Springer-Verlag, 1985.

[GLSW]   Gotay, M.; Lashof, R.; Sniatycki, J.; and Weinstein, A. Closed forms on
            symplectic fiber bundles. *Comm. Math. Helv.* **58**, no. 4 (1983), 617–21.

[GN]   Gotay, M., and Nester, J. M. Generalized constraint algorithm and
            spacial presymplectic manifolds. In *Geometric Methods in
            Mathematical Physics*, G. Kaiser and J. E. Marsden, eds., Springer
            lecture notes in math. 775, 78–104. Berlin: Springer-Verlag, 1980.

[Gr]   Gromov, M. Pseudo-holomorphic curves in symplectic manifolds.
            Invent. Math. **82** (1985), 307–47.

[GK]   Grossberg, M., and Karshon, Y. Bott towers, complete integrability, and
            the extended character of representations. *Duke Math. J.* **76** (1994),
            23–58.

[Gil1]   Guillemin, V. *Moment Maps and Combinatorial Invariants of
            Hamiltonian T^n-spaces*. Boston: Birkhäuser, 1994.

[Gil2]   Guillemin, V. Reduced phase spaces and Riemann–Roch. In *Lie Theory
            and Geometry: In Honor of Bertram Kostant*. J.-L. Brylinski et al., eds.
            Boston: Birkhäuser, 1994.

[GC]   Guillemin, V., and Canas, A. On the Konstant multiplicity formula for group actions with nonisolated fixed points. Preprint, MIT, Cambridge, MA, 1995.

[GS1]   Guillemin, V., and Sternberg, S. *Geometric Asymptotics*. Providence, RI: American Mathematical Society, 1977.

[GS2]   Guillemin, V., and Sternberg, S. Convexity properties of the moment mapping. *Invent. Math.* **67** (1982), 491–513.

[GS3]   Guillemin, V., and Sternberg, S. Geometric quantization and multiplicities of group representations. *Invent. Math.* **67** (1982), 515–38.

[GS4]   Guillemin, V., and Sternberg, S. Homogenous quantization and multiplicities of group representations. *J. Func. Anal.* **47** (1982), 344–80.

[GS5]   Guillemin, V., and Sternberg, S. The Gelfand–Cetlin system and quantization of the complex flag manifolds. *J. Func. Analysis*, **52** (1983), 106–28.

[GS6]   Guillemin, V., and Sternberg, S. The Frobenius reciprocity theorem from a symplectic point of view. In *Non-linear Partial Differential Operators and Quantization Procedures: Proceedings of a Workshop Held at Clausthal, Federal Republic of Germany, 1981*, S. I. Andersson and H.-D. Doebner, eds., Lecture notes in mathematics 1037, 242–56. Berlin: Springer-Verlag, 1983.

[GS7]   Guillemin, V., and Sternberg, S. Multiplicity-free spaces. *J. Diff. Geom.* **19** (1984), 31–56.

[GS8]   Guillemin, V., and Sternberg, S. *Symplectic Techniques in Physics*. Cambridge: Cambridge University Press, 1984.

[GS9]   Guillemin, V., and Sternberg, S. Birational equivalence in symplectic category. *Invent. Math.* **97** (1989), 485–522.

[GS10]   Guillemin, V., and Sternberg, S. *Variations on a Theme by Kepler*. Providence, RI: American Mathematical Society, 1990.

[Han]   Hannay, J. H. Angle variable holonomy in adiabatic excursions of an integrable Hamiltonian. *J. Phys. A* **18** (1985), 221–30.

[Hec]   Heckman, G. Projections of orbits and asymptotic behavior of multiplicities for compact Lie groups. Ph.D. thesis, University of Leiden, 1980.

[Hel]   Helgason, S. *Differential Geometry, Lie Groups, and Symmetric spaces*. New York: Academic Press, 1978.

[Her]   Hermann, R. On the differential geometry of foliations. *Ann. Math.* **72** (1960), 445–57.

[HPS]   Hogreve, H.; Potthoff, J.; and Schrader, R. Classical limits for quantum particles in external Yang–Mills potentials. *Comm. Math. Phys.* **91** , no.4 (1983), 573–98.

[Hör]   Hörmander, L. *The Analysis of Linear Partial Differential Operators I*. New York: Springer-Verlag, 1983.

[Hum]   Humphreys, J. E. *Introduction to Lie Algebras and Representation Theory*. New York: Springer-Verlag, 1972.

[I]   Iwai, T. On reduction of the four-dimensional harmonic oscillator. *J. Math. Phys.* **22** (1981), 1628.

[J]   Jacobson, N. *Lie Algebras*. New York: Dover, 1979.

[JK1]   Jeffrey, L., and Kirwan, F. Localization for non-abelian group actions. *Topology* **34** (1995), 291–327.

[JK2]   Jeffrey, L., and Kirwan, F. On localization and Riemann-Roch numbers
            for symplectic quotients. Preprint, alg.-geom. 9506007, 1995.

[Kac]   Kac, V. *Infinite Dimensional Lie Algebras*. Cambridge: Cambridge
            University Press, 1990.

[Kal]   Kalkman, J. Cohomology rings of symplectic quotients. *J. reine angew.
            Math.* **458** (1995), 37–52.

[KK]   Kantor, J. M., and Khovanskii, A. Integral points in convex polyhedra,
            combinatorial Riemann–Roch theorem and generalized MacLaurin
            formula. *Pub. Math. IHES* (1992), 932–7.

[KT]   Karshon, Y., and Tolman, S. The moment map and line bundles over
            pre-symplectic toric manifolds. *J. Diff. Geom.* **38** (1993), 465–84.

[Kaw]   Kawakubo, K. *The Theory of Transformation Groups*. Oxford: Oxford
            University Press, 1991.

[KKS]   Kazhdan, D.; Kostant, B.; and Sternberg, S. Hamiltonian group actions
            and dynamical systems of Calogero type. *Comm. Pure Appl. Math.* **31**
            (1978), 481–507.

[KP]   Khovanskii, A., and Pukhilov, S. Théorème de Riemann–Roch pour les
            intégrales et les sommes de quasi-polynomes sur les polyédres virtuel.
            Algebra i analyz **4** (1992), 188–216.

[K1]   Kostant, B. Orbits, symplectic structures and representation theory. In
            *Proceedings, US–Japan Seminar in Differential Geometry, Kyoto
            (1965)*. Tokyo: Nippon Hyoronsha, Tokyo, 1966.

[K2]   Kostant, B. Quantization and unitary representations. In *Lectures in
            Modern Analysis and Applications III*, C. T. Taam, ed., Lecture notes in
            mathematics 170, Berlin: Springer-Verlag, 1970.

[K3]   Kostant, B. On convexity, the Weyl group and the Iwasawa
            decomposition. *Ann. Sci. Ec. Norm. Sup.* **6** (1973), 413–55.

[KS]   Kostant, B., and Sternberg, S. Symplectic reduction, BRS cohomology
            and infinite dimensional Clifford algebras. *Ann. Phys.* **176** (1987),
            49–113.

[L]   Lerman, E. Symplectic cuts. *Math. Research Lett.* **2** (1995), 247–58.

[LT]   Lerman, E., and Tolman, S. Symplectic toric orbifolds. Preprint, MIT,
            Cambridge, MA, dg-ga/9412005, 1994.

[McD]   McDuff, D. Examples of simply-connected non-Kälerian manifolds.
            J. Diff. Geom. **20** (1984), 267–77.

[MMR]   Marsden, J. E.; Montgomery, R.; and Ratiu, T. Reduction, symmetry
            and phases in mechanics. *A.M.S. Memoirs* **436** (1988).

[MW]   Martin, S., and Weitsman, J. On a conjecture of Guillemin–Sternberg.
            In preparation.

[Mein1]   Meinrenken, E. On Riemann–Roch formulas for multiplicities.
            J. A.M.S., in press.

[Mein2]   Meinrenken, E. Symplectic surgery and $Spin^c$-Dirac operator.
            Advances in Math., in press, dg-ga/9504002.

[MF]   Mishchenko, A. S., and Fomenko, A. T. A generalized Liouville
            method for integration of Hamiltonian systems. *Funk. Analiz Pril.* **12**
            (1978), 46–56.

[Ml]   Mladenov, I. M. Geometric quantization of the MIC–Kepler problem
            via extension of the phase space. *Ann. Inst. H. Poincaré Phys. Theor.* **50**
            (1989), 183–91.

[M1]   Montgomery, R. Canonical formulation of a classical particle in a
        Yang–Mills field and Wong's equations. *Lett. Math. Phys.* **8** (1984),
        59–67.

[M2]   Montgomery, R. The bundle picture in mechanics. Ph.D. thesis, UC
        Berkeley, 1984.

[M3]   Montgomery, R. The connection whose holonomy is the classical
        adiabatic angles of Hannay and Berry and its generalizations to the
        non-integrable case. *Comm. Math. Phys.* **120** (1988), 269–94.

[Mo]   Moser, J. On the volume element on a manifold. *Trans. AMS* **120**
        (1965), 286–94.

[O'N]   O'Neill, B. The fundamental equations of a submersion. *Mich. J. Math.*
        13 (1966), 459–69.

[Pom]   Pommersheim, J. Toric varieties, lattice points and Dedekind sums.
        Math. Ann. **295** (1993), 1–24.

[P]   Pukanszky, L. Unitary representations of solvable groups. *Ann. Sci.*
        Ecole Normale Sup. **4** (1971), 457–608.

[R]   Rawnsley, J. Representations of a semi-direct product by quantization.
        Math. Proc. Cambridge Phil. Soc. **78**, no. 2 (1975), 345–50.

[Sa]   Satake, I. The Gauss–Bonnet theorem for V-manifolds. *J. Math. Soc.*
        Japan **9** (1957), 464–92.

[Sch]   Scheunert, M. *The Theory of Lie Superalgebras: An Introduction.*
        Lecture notes in mathematics 716. Berlin: Springer-Verlag, 1979.

[SL]   Sjamaar, R., and Lerman, E. Stratified symplectic spaces and reduction.
        Ann. Math. **134** (1991), 375–422.

[St]   Steenrod, N. E. *Topology of Fiber Bundles.* Princeton, NJ: Princeton
        University Press, 1951.

[S1]   Sternberg, S. Symplectic homogeneous spaces. *Trans. A.M.S.* **212**
        (1975), 113–30.

[S2]   Sternberg, S. On minimal coupling, and the symplectic mechanics of a
        classical particle in the presence of a Yang–Mills field. *Proc. Nat.*
        Acad. Sci. USA **74** (1977), 5253–4.

[Ver]   Vergne, M. Multiplicities formulas for geometric quantization, I, II.
        Preprints, Ecole Normale Supérieure, Paris, 1994.

[W1]   Weinstein, A. *Lectures on Symplectic Manifolds.* CBMS Lecture Notes
        29. Providence, RI: A.M.S., 1977.

[W2]   Weinstein, A. A universal phase space for particles in Yang–Mills
        fields. *Lett. Math. Phys.* **2** (1978), 417–20.

[W3]   Weinstein, A. Fat bundles and symplectic manifolds. *Adv. Math.* **37**
        (1980), 239–50.

[W4]   Weinstein, A. Connections of Berry and Hannay type for moving
        Lagrangian submanifolds. *Adv. Math.* **82** (1990), 133–59.

[Weit] Weitsman, J. A Duistermaat–Heckman formula for symplectic circle
        actions. *Int. Math. Research Notices* (1993), 309–12.

[Zel]   Želobenko, D. P. *Compact Lie Groups and their Representations.*
        Transl. of Math. Monographs 40, Providence, RI: A.M.S., 1973.

Index